Additional Praise for *Seeing the Unseen*

"The authors do a masterful job of demystifying the key factors driving the rise of world-class technology and entrepreneurship from China. Through first-hand experience as entrepreneurs, strategy leaders, and investors from China doing business globally, they offer insights and practical approaches for entrepreneurs everywhere to better compete both in China and abroad."
—Blake Larson, global tech entrepreneur

"A standout in the multitude of works explaining what Chinese technology companies do and how, *Seeing the Unseen* expertly explains their 'why' and offers a practical framework for collaboration—or competition—with Chinese founders as they take on the world. Interwoven with well-placed forays into China's rich history, a delight to read."
—Dmitry Levit, partner, Cento Ventures

"*Seeing the Unseen* discusses the past, present, and future of Chinese tech firms. It discusses why and how giants rose in the past, the challenges faced today in the global arena, and what we can expect in the future. The book provides deep insights to anyone who wants to know about Tech China and rising players from emerging markets."
—Feng Zhu, professor of business administration, Harvard Business School

"*Seeing the Unseen* deciphers much of the philosophy in action on how entrepreneurs think and how they make crucial decisions. There are many gold nuggets in understanding why key events in Chinese entrepreneur history happened the way they did."
—Jianfeng Lu, chairman, Wiz Holdings (wiz.ai), serial entrepreneur and investor

"My great-grandparents once told us to keep on learning all the way to China, as the rich history there offers a lot of lessons for our work and life. This book, which is very detailed and informative, shows just that. It is a strong combination of in-depth research and life experience of the authors in the tech industry. A must-read!"
—Kameswara Natakusumah, country leader and president director, Willis Tower Watson Indonesia

"For those who are interested in or embedded in the ecosystem of Chinese tech firms, *Seeing the Unseen* is a delightful and thorough guide full of information and insights. Reading it is not only a must, but a wonderful journey led by Guoli and Jianggan. Both have a profound yet objective understanding of the history and leadership of Chinese tech giants and how they impact the world."

—Randolph Hsu, founder of Ondine Capital

"For anyone interested in Chinese tech companies, don't walk, but run, to get this book. As deeply embedded experts in the Chinese tech context, the authors raise the curtain on China and its rising tech giants."

—Riitta Katila, W. M. Keck Sr. chaired professor of management science, and research director of the Stanford Technology Ventures Program, Stanford University

"*Seeing the Unseen* explains many of the enigmas that I experienced as an expat while working and living in Asia. If I could go back in time, I'd better understand why Chinese tech companies operate the way they do, and how to better work with them. Fully recommended reading!"

—Rodrigo Becerra Mizuno, chief technology officer, Televisa Univisión

Seeing the Unseen

Behind Chinese Tech Giants'
Global Venturing

GUOLI CHEN
JIANGGAN LI

WILEY

For general information on our other products and services or for technical support, please contact our Customer Care Department within the United States at (800) 762-2974, outside the United States at (317) 572-3993 or fax (317) 572-4002.

Wiley also publishes its books in a variety of electronic formats. Some content that appears in print may not be available in electronic formats. For more information about Wiley products, visit our web site at www.wiley.com.

Library of Congress Cataloging-in-Publication Data is Available:

ISBN 9781119885832 (Hardback)
ISBN 9781119885856 (ePDF)
ISBN 9781119885849 (ePub)

Cover Design: Wiley
Cover Image: © hudiemm/Shutterstock

SKY10035538_080122

This book is dedicated to my parents, Chen Xishi and Luo Yuzhen, and family members for their constant love and support.
—Guoli Chen

This book is dedicated to my dad, Li Yue; my mum, Jiang Meijuan; my uncles, Jiang Jian and Jiang Guo; and my aunts, Li Yinyan and Li Xiaoyan; as well as my extended family. Your support and encouragement over the years allowed me to be curious, explore, and ultimately share Chinese stories with the world.
—Jianggan Li

Contents

PART III: RESTEERING THE WHEEL

Preface

WE HAVE BEEN WORKING with each other through an INSEAD course, "China Strategy," where participants and students increasingly took an interest in an emerging phenomenon: Chinese tech and internet companies expanding out of China. Through interactions at the course, we began to formulate a clear picture about what is really happening behind these waves of global expansion. The seed of co-authoring a book to put the strategies, stories, and lessons learned of Chinese tech firms' global journeys had been planted.

An increasing number of Chinese firms stepped out of China and became more active in the international market to pursue their global ambitions. For instance, in 2014, Tencent launched a large TV campaign across the world featuring famous footballer Lionel Messi promoting WeChat; in 2016, Alibaba acquired Lazada, then Southeast Asia's leading ecommerce platform; the following year, Souq, the Middle East's largest ecommerce platform, was sold to Amazon at a steep 45% discount—one of the reasons was the threat from a previously unheard-of Chinese ecommerce company called JollyChic; in 2019, Transsion, a Chinese smartphone player that became the largest (by sales) in Africa, went public; in 2020, TikTok became the biggest national security threat to the United States in the eyes of Donald Trump; in April 2022, China-based fast-fashion ecommerce player SheIn was valued at $100 billion in a funding round, worth more than H&M and Zara combined.

There are many more less-known companies, as well as companies such as Singapore-based Sea Group, founded by the global Chinese community and inspired by business models in China, that are making their presence felt in different parts of the world.

We have been receiving more and more questions from various audiences and clients from outside China on this phenomenon, its potential impact, and how different parts of the business ecosystem should respond to it. Senior corporate leaders participating in programs at INSEAD's France campus debated about the data privacy implications of Alibaba and its affiliate companies, including Ant Group, whose IPO was shelved by the Chinese government in 2020; executives of Fortune 500 firms raised questions about Chinese leadership styles and the myth of sudden rise, and tried to understand how the change of policies influences the competition and cooperation between Chinese and US tech firms; the scion of a large media conglomerate in the Philippines inquired how they could revamp their media assets and learn from Toutiao, the original success of TikTok parent ByteDance.

And across Southeast Asia, South Asia, and Africa, financial regulators have been working very hard to figure out how to regulate the array of fintech companies copying Chinese models, employing Chinese technical experts, and being invested in by Chinese groups.

We have also received a large number of enquiries from entrepreneurs, company leaders, and investors from China looking to expand into foreign markets. Where to start, what to prepare, whom to hire locally or send as expats, how much resources to allocate—for these Chinese tech leaders who are used to fighting big battles in a large single-country market, the variety and fragmentation of global markets are daunting.

The more discussion and debates we had on these questions with various parties, the more we felt that we needed to do something. The COVID-19 pandemic, which grounded us as well as our ecosystem interlocuters, became the catalyst for us to commit our efforts to synthesize all these observations, thoughts, and reflections into writing.

We hope that you find useful insights in this book—no matter if you're working with, competing against, joining, or regulating these Chinese tech firms (and those inspired by them), or being part of their global journey.

We are also embedding some traditional Chinese idioms and phrases that might help explain some actions and situations—we try our best to make them relevant and easy to understand.

As this is a fast-evolving sector in which changes happen every month, especially in the current global geopolitical environment we are living in, we look to update you with the latest happenings, case studies, and reflections as they happen. Subscribe to our newsletter at http://www.poplstrategy.com/. You can also reach us through the website if you would like to discuss these topics with us.

Acknowledgments

WHILE WE HAD THE idea for this book for a long time, our jet-setting lifestyles were prohibitive. When the COVID-19 pandemic and the lockdowns were enforced in Singapore, we decided to kickstart the project. To our pleasant surprise, so many friends and colleagues have come in to help, offering experiences, insights, perspectives, and reflections. Many have introduced us to their circle of friends and (ex)colleagues for us to get a deeper understanding of particular companies, episodes, or people.

Specifically, in formulating and compiling this book, we interviewed many investors, entrepreneurs, and executives of major companies, who have participated in the process of Chinese companies venturing overseas. In over 300 hours of interviews, we heard detailed accounts as well as reflections of journeys undertaken by many companies, but, more importantly, many more pioneers who led and participated in the process.

We are grateful that a few especially influential tech leaders shared their perspective with us, and allowed us to use their quotes, although all of them requested that they remain anonymous.

Andy Li, Canary Zhang, Dan Hu, Diane Weng, Harry Xue, Jonathan Zhong, Nick Duan, Qi Zhai, Shaolin Zheng, Tao Tian, Thomas Shi, Vincent Yang, Yongqing Li, Yun Zhang, Zhen Zhang, and many others, who prefer to remain anonymous, have shared their personal experiences and reflections.

We would like to thank Yorlin Ng, chief operating officer of Momentum Works, for sharing and reflecting real-life leadership and people lessons, which became valuable insights for the book; Steven Peng, visiting PhD candidate at INSEAD, and Deshui Yu, project manager at Momentum Works, who lent a big hand in attending interviews, organizing all the notes, and deliberating on the key points with us over many weeks.

We would also like to thank other colleagues at Momentum Works, without whose help this book would have not been possible: Crystal Yu, Nanette Litya, and Nurina Fhareza, for their respective experiences working with Chinese tech companies in and outside China; Aishwarya Valliappan, Vion

Yau, Brandon Yee, and Yi Hu for their specific work at various stages of this project. We are grateful for colleagues at INSEAD, students in the classrooms, and participants in research seminars and conferences. We are also indebted to support from Quy Huy and INSEAD's China Initiative, which provides sources in developing China-related pedagogical cases.

We are sure we have neglected many more who have helped in this book in one way or another. Also, because of the limited space in this book, we could not accommodate many experiences from many other companies that were shared to us—we will continue to share some of these through other channels, including our blog, http://www.poplstrategy.com/.

Finally, we would like to thank all the investors, entrepreneurs, and executives with whom we have discussed the topic of Chinese tech venturing overseas over the years. Without your collective insights, we would not dare to touch such a big, complex, and fast-evolving topic.

Guoli Chen and Jianggan Li
May 5, 2022
Singapore

About the Authors

Dr. Guoli Chen is a professor of strategy at INSEAD. He received his PhD in strategic management from the Pennsylvania State University. Guoli's research interest focuses on strategic leadership and corporate governance in the context of mergers and acquisitions, IPOs, innovation, globalization, corporate renewal, and sustainability. He has published more than 20 academic papers in top management journals, and his articles and opinions also appear in many newspapers and magazines. Guoli has served in various leadership roles in the Strategic Management Society (SMS) and the Academy of Management (AoM). Guoli is a senior editor of the *Management and Organization Review* (*MOR*), the most impactful academic journal on China-related research. He is an expert in China strategy and has also published several pedagogical cases on Chinese companies, such as "Huawei's Smartphone Strategy," "Uber vs. Didi," "Ant Financial and Tencent," "SheIn vs. Zara," and "TikTok and Kuaishou."

Jianggan Li is the founder and CEO of Momentum Works, a Singapore-headquartered venture outfit with a global emerging market focus. Before launching Momentum Works in 2016, he built and scaled internet and technology companies across Southeast Asia at Rocket Internet. Specifically, he cofounded Easy Taxi in Asia, and served as managing director of Foodpanda. In addition to Southeast Asia, he has lived and worked in Greater China, India, Europe, and Latin America. Now, in addition to building ventures, Jianggan leads a team on producing insights into new fast-moving industries that help individuals, investors, and corporations connect the dots and make an impact. He speaks at many conferences and corporate events.

Jianggan holds an MBA from INSEAD and a degree in computer engineering from Nanyang Technological University. Apart from English and his native Mandarin, he is also fluent in French and conversational in Cantonese and Spanish.

Introduction

I N MAY 2020, KEVIN Mayer, a seasoned media executive, resigned from the Walt Disney Company to join ByteDance, raising a lot of eyebrows.

For the first time, an established Chinese tech company managed to convince such a high-profile American executive to join—in the executive's home market. Even more impressive, Mayer's new role, CEO of TikTok and COO of its parent company ByteDance, made him the first American executive given the power to run a major and most critical business unit of a Chinese internet company.

He was obviously qualified: at Disney he was in charge of streaming businesses, international channels, advertising sales, and distribution, as chairman of Walt Disney Direct-to-Consumer & International (DTCI). There were rumors that his departure was because Bob Iger, the outgoing CEO of Disney, named Bob Chapek instead of Mayer as the successor of the Disney empire.

Commentators were, however, split. Some cheered this as a new era, where Chinese tech companies can tap into top global talent; others were more skeptical, citing the cultural fit that has crippled many other foreign executives (mostly of lower ranks) in Chinese tech companies. Either way, people agreed on ByteDance's boldness in international expansion, taking a step that East Asian companies rarely take.

Barely three months later, Mayer announced his resignation from his roles at ByteDance, leaving the company.

Even from the outside, the three months looked like a wild roller-coaster ride. The Trump administration gave ByteDance an ultimatum to either shut down TikTok or sell it to an American company. Intensive discussions happened with multiple acquirers and potential partners, including Microsoft, Oracle, and Walmart. Pressure probably mounted also from investors, internal teams, and the Chinese government. Only Mayer himself knows what he went through.

Some wondered, if Trump had not forced a sale, would Kevin have survived or thrived under Zhang Yiming, the founder of ByteDance, one of the most valuable private tech companies originated from China which is well known globally for its crown jewel app TikTok?

ByteDance, together with many other Chinese tech firms such as Alibaba, Tencent, Baidu, Xiaomi, Pinduoduo, and Meituan, have become more and more regularly hitting the news headlines in the past decade, not only because of their extraordinary growth, recent upheaval (due to China's tech crackdown and US–China geopolitics), but also their increasing international presence and market influence.

From copycats of Silicon Valley's trendy ideas to increasingly making their global presence known, Chinese tech and internet companies have come a long way, in a very short period of time.

However, beyond the media headlines and financial reporting from a few big-listed companies, little is known about China's tech firms: Who are they? How did they become what they are today? What are their real competitive advantages? What are their global ambitions? Can they achieve these ambitions?

These questions frequently surface in our discussions with various tech stakeholders across continents, and for the right reasons: companies need to figure out whether to compete, or collaborate, with these Chinese firms, and how. Experienced talent is getting recruiter calls from these firms, but the talent has heard little about these firms' culture or their financial prospects. Regulators need to figure out how to deal with hundreds of permutations of business models, some never seen outside China.

Media and analyst reports can also be misleading. For example, Didi was penalized because of its monopolistic behavior (the main concern is cybersecurity). Pinduoduo, which really started as a gamified version of Taobao, is often portrayed as "social ecommerce"; Ant Group's credit-scoring system (Zhima Credit) is credited with helping the company increase the performance of its lending products (it's not a full picture).

In this book, we seek to first give the readers *an accurate, concise understanding of Chinese tech companies*, and then, more importantly, we want to focus on their global expansion efforts. We have summarized the key lessons and future propositions into a *POP-Leadership* (POP stands for product, organization, and people) framework, which will be discussed later.

The journey has not been smooth for most of the companies, with failures more common than successes. However, it is important to note that as a whole, Chinese tech firms learn very fast, from their humble beginning as copycats.

WHO SHOULD READ THIS BOOK?

Who will find this book interesting? In a market that has only recently emerged, these Chinese tech giants are influential, yet understudied. While abundant media coverage has been produced about some of these companies, little information exists regarding their leadership thinking, strategy, organization, strengths, and weaknesses. Now is such a crucial time—with potential opportunities yet a lack of information—for *regulators, potential partners, competitors, suppliers, customers, experienced professionals, and other stakeholders* to better understand how these companies operate.

We believe you will find the book useful if you are any of the following:

- A senior executive leading a company to compete or cooperate with Chinese companies, in or outside China.
- A business owner embedded in the supply chain or ecosystem led by Chinese tech firms.
- A budding entrepreneur who would like to replicate the business models outside China.
- A consultant who advises on the strategies Chinese high-tech firms have used to rapidly grow over the past decades.
- A professional receiving a call from a headhunter who is going to offer you a generous package to work for a Chinese tech firm in Phnom Penh.
- A researcher aiming to write a report with a specific focus on the Chinese tech industry.
- A government official who would like to formulate policies to attract and manage foreign direct investment (FDI) and new business models from China (and yet are bewildered by the recent regulatory crackdowns on big techs in China).

This is because, whether you are of Chinese background or not, a tech company or an entrepreneur, a researcher or business enthusiast, this book aims to fill a gap in understanding how Chinese companies create, compete, and venture into the international landscape. In addition to these companies, the global Chinese community has already been copying models from their cousins in China and achieving successes outside China—Singapore-based Sea Group (market cap: $159 billion on December 1, 2021) is a case in point.

 ## SCOPE OF THIS BOOK: WHAT ARE TECH/INTERNET COMPANIES?

We use the terms "high-tech" and "internet" firms loosely, but what are internet companies? Before we get into more details about our framework, we would like to define the term and set the scope of this book.

A narrow definition of internet firms is those companies that rely on the internet (more recently the mobile internet) as the main distribution channel to reach out to their base of customers. A broader definition involves companies using (mobile) internet as a key differentiator that separates them from other firms serving the same demand (think about ecommerce as compared to traditional retail); it also includes firms that serve these companies using internet-enabled technologies, such as logistics, supply chain management, credit assessment, and cloud computing.

Internet companies generate revenues and profits primarily through:

1. Fees or commissions for facilitating transactions/sales (marketplace model)
2. Advertising
3. Charges for providing a piece of infrastructure or service (e.g., logistics, credit assessment, cloud)
4. Direct online sales
5. Financial transaction fees and interests
6. Ecommerce live-streaming
7. Payment for gaming/content and other in-app purchases

Among these monetization methods, 1 and 2 are mainly levied on businesses; 3, 4, and 5 are levied on businesses and/or consumers; 6 and 7 are usually levied on consumers.

Because of their nature as disruptors in a very fast-growing field, large internet companies tend to be strategically aggressive in fending off potential disruptors. One method is by expanding their customer offerings and providing additional value to attract more customers and retain existing customers.

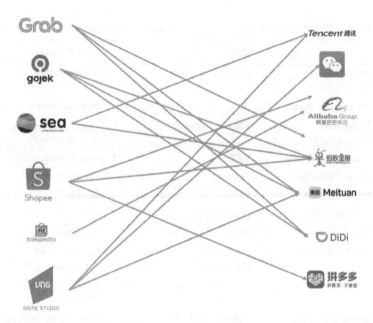

FIGURE I.1   The matching between companies in emerging markets and Chinese internet firms
Source: Momentum Works.

This leads to what we call the "super app" strategy,[1] as demonstrated by WeChat, Meituan, and Alipay: consumers can shop, order food delivery, make payments, and access municipal and other services, all on the same app. The strategy helps the firms deepen their moat and fend off potential competitors. It also helps the companies extract more value from their customers—this is especially the case in China where the average customer value is still considerably lower compared to their US peers.

Certain US companies are also taking the same approach, albeit with less urgency; companies in emerging markets tend to copy the Chinese model more, as the market realities they face resemble those of China more (Figure I.1).

The definition of internet companies given at the start of this section helps us generate a table of Chinese versus US internet firms in different domains of businesses (Table I.1). The purpose is not to show a comprehensive list, but to give an idea of what types of firms we focus on in this book.

TABLE I.1   Chinese Internet Firms versus US Internet Firms

		China	United States
1	Shopping—Marketplace	Alibaba/Pinduoduo	Amazon/eBay
2	Shopping—Retail	JD	Walmart/Amazon
3	Shopping—Nonplatform	Youzan, Weimob	Shopify (Canadian), BigCommerce
4	Shopping/Vertical	VIPS	Etsy, Chewy, Wayfair
5	Social/Messaging	Tencent	Facebook
6	Search engine	Baidu	Google
7	Ride hailing	Didi, Meituan	Uber/Lyft
8	Food delivery	Meituan, Alibaba	Doordash, Uber
9	Short video	Tencent/Kuaishou/Douyin (Bytedance)	None—TikTok dominates
10	Online travel	Ctrip.com, Meituan, Alibaba	Booking.com, Expedia, TripAdvisor
11	Music streaming	Tencent, NetEase	Spotify (Swedish), Apple
12	Digital payment/wallet	Ant Group, Tencent	Square/Paypal/Venmo
13	Smartphones	Xiaomi/Huawei/Oppo/Vivo/Oneplus/Realme/Transsion	Apple
14	Video streaming	Iqiyi/Tencent	Netflix/Amazon/Disney+
15	Property	Beike	Zillow/Opendoor
16	Used car	Guazi	Carvana
17	Intercity logistics	Lalamove	None (Lalamove expanding)

POP-LEADERSHIP

Searching for a global market is not unique to Chinese firms. In their overseas journey, Chinese firms face some of the same or similar challenges that US, European, Japanese, and Korean companies faced in the past. However, the key differences faced by the current wave of Chinese internet companies include, but are not limited to, the following:

■ The pace is much faster, compared to traditional industry such as Unilever investing decades in cultivating the Chinese market, or Toyota investing in Indonesia.

- Chinese firms have a large domestic market (compared to Japanese, Korean, or even European markets), which is a double-edged sword for their globalizing efforts.
- Chinese internet firms have developed unique capabilities that specifically fit the Chinese market but may not fit other markets.
- Chinese firms do not have a natural advantage like US tech giants when they expand globally.

These unique features of Chinese internet firms in their overseas ventures can challenge our thinking on the global strategy and implementation.

Keeping in mind the commonalities and differences of Chinese internet firms and other companies, we are using the *POP-Leadership* framework to describe and help you understand the key aspects of Chinese companies' overseas ventures (Figure I.2).

Leadership

Leadership is in the central place, because leaders make decisions on behalf of their firms, and their actions (and inactions) will be amplified through the organizational system. Leaders set the ambition and direction for their firms, make ultimate decisions to advance or retreat, allocate, or cut resources in different markets. We have seen firms become legendary because of their leaders'

FIGURE I.2 POP-Leadership

inspiring vision and determination. We also witnessed many other firms collapse because of their leaders' overconfidence, vacillation, and misjudgment at critical stages.

To understand a firm's strategy, we first need to study its strategist. Indeed, leaders provide a shortcut for us to comprehend or even foresee the next strategic move, as decades of research have tried to understand how top executives' experiences, cognition, emotions, and psychological attributes influence organizational choices and strategic outcomes.[2]

Leaders of Chinese tech firms grow up in their unique environment, and thus have different imprinting effects on the organizations they create. Despite its relatively short history, the Chinese internet industry has experienced different generations of tech leaders, from early pioneers such as William Ding (founder of NetEase.com) and Charles Zhang (founder of Sohu.com) to recent iconic figures such as Yiming Zhang and Colin Huang. The newer generation of tech leaders has a more international mindset from the beginning.

However, overseas expansion for tech firms that have established a presence in China is a second entrepreneurial journey during which their leaders are going to face new challenges in a different environment unfamiliar to them. Meanwhile, these leaders have to deal with domestic Chinese businesses that are hypercompetitive, too.

Thus, in addition to being thoughtful strategists, relentless executioners, and resilient survivors, which makes them who they are today, what more is needed for the global expansion of tomorrow? We draw from prior experience of Chinese tech firms to unveil the leadership considerations in overseas venturing. For instance, how does these leaders' mental space switch between two markets (domestic versus international)? What resources and managerial attention will need to be invested or allocated in new territories? How do leaders navigate their firms through these complex situations?

People

People is the second element we discuss in the framework. Undoubtedly, people who make things happen are the core of the organization, not only in executing strategy, but also in formulating strategy and in adaptation (i.e., in the bottom-up process of responding to market feedback).

Chinese internet firms have a unique trajectory in the people they can hire, from a very humble beginning where no one really wants to join (think about the earlier days of Alibaba and Tencent) to the most attractive places where all the young talent wishes to be enlisted. Chinese tech firms have recently started aggressive global hiring of senior executives.

Experienced seniors are those who matter the most, but are also the hardest to manage in the organization. There are different issues for those veterans internally promoted through the ranks and those hired externally, especially in overseas markets. How do you get strong lieutenants to operate and tackle the challenges of global expansion? What can you learn from the people and HR systems that currently function well in China? More importantly, how do you fill the gap of people issues and crack the code in overseas expansion?

Organization

Organization is the third element in our framework. It addresses the issue of how to group people together to achieve a common goal. Doing this is never an easy task, especially when these people are strangers, and the goals need to be figured out along the way in a fast-paced tech environment. The challenge is further amplified when firms grow bigger, absorbing more people whose experiences, life goals, and ways to approach work are not necessarily fully aligned with each other.

Chinese internet firms and their leaders have designed their own systems to address the organizational issues in their journey to become the unicorns and then the tech giants. In other words, it is a selection process of survivors: firms that cannot address the organization issue have disappeared or become irrelevant in today's business landscape.

To keep organizational dynamism and agility, Chinese tech firms have developed several tactics, such as continuous organizational restructuring, the rotation of executives, and intense internal competition. However, what make these firms dynamic and agile to adapt to the changing environment in China will also create problems in their overseas businesses.

We discuss how the fundamental issues in the organization—decision rights, information communication, and resource allocations—arise if firms simply copy and paste the structure that works well in China to overseas markets. Again, drawing from the experience of many Chinese tech firms, we suggest several questions related to organization for leaders to consider in their overseas venturing.

Product

Product is the last, but not the least, element in our framework. The implications of leadership, people, and organization have to be translated into products (both tangible offerings or intangible services) to engage with clients, to generate and capture value.

In overseas expansion, product is related to a series of decisions that include three Ws and one H: what to offer, when to enter, where to enter, and how to enter foreign markets.

The Chinese internet is a relatively closed ecosystem and Chinese users have different behaviors that have also been evolving over the past decades because of high economic growth and new technology adoption. Such an environment breeds different products, such as supper apps.

Thus, in their overseas venturing, leaders need to decide what to adapt to the local market (i.e., the trade-off between localization and standardization in global businesses). We also discuss various perspectives of key questions of where, when, and how to enter. Proposing a one-size-fits-all solution is difficult; instead, we lay out key considerations in product-related decisions in overseas venturing.

To summarize, POP-Leadership covers leadership, people, organization, and product, which synthesizes key themes of successful overseas ventures, as well as the key challenges along the way. The framework facilitates strategists to dissect the ambitious goal into manageable pieces, guiding them to think systematically to identify the potential pitfalls and the key bottlenecks when they consider their own global expansion.

 BOOK STRUCTURE

In **Part One**, which includes three chapters, we first introduce who these Chinese tech firms are and how they copy and innovate, compete, and grow. This information will provide you with a better understanding of these emergent players in the global arena.

In **Part Two**, which includes four chapters, we get into the details of each parameter of the POP-leadership framework, with Chapter 4 focusing on leadership, Chapter 5 on people, Chapter 6 on organization, and Chapter 7 on product. Although we illustrate the parameters one by one, all of them are essentially interconnected.

In **Part Three**, we reflect on the latest happenings in the China-US geopolitical tussle, the regulatory crackdowns in China in 2021, and the rise of Chinese-inspired companies globally, particularly in Southeast Asia. We also take the perspectives of the local players to understand the potential spillover effect triggered by Chinese tech firms' overseas investments.

Finally, we present Chinese tech firms' global map and a picture of cross-pollination of global emerging markets. We believe it is still in the early stage, and expect a few full blossoms in the future, despite the enormous challenges ahead.

NOTES

1. G. Chen and M. Trocha, "Super Apps: How to Create a Mass Market of One," INSEAD Knowledge, October 2, 2019.
2. S. Finkelstein, D. Hambrick, and A. Cannella, *Strategic Leadership: Theory and Research on Executives, Top Management Teams, and Boards* (New York: Oxford University Press, 2009).

PART ONE

Understanding Chinese Tech Companies

CHAPTER ONE

The Shark versus the Crocodile

eBay may be a shark in the ocean, but I am a crocodile in the Yangtze River. If we fight in the ocean, we lose—but if we fight in the river, we win.

—Jack Ma, founder of Alibaba Group, 2003

I N 2003, WHEN EBAY, a leading US ecommerce platform, entered China, most analysts and investors were bearish about a new ecommerce venture a young company called Alibaba had just launched: Taobao. After all, Taobao's initial investment of RMB 100 million was nothing compared to eBay's market cap of more than US$30 billion, let alone the ecommerce know-how, firm reputation, and attractiveness to good talent.

Jack Ma, the founder of Alibaba, was determined to push through. He said that a crocodile will beat the shark in a fight in the river.

A year later, in 2004, Taobao was ahead of eBay in most metrics in China, defying the predictions of analysts and investors. The same year, another giant shark entered the Yangtze: Amazon acquired Joyo to enter China.

Since then, Amazon itself has grown from a $15 billion company to a $1.8 trillion behemoth. However, it did not win China as it had hoped. In 2019, faced with a less than 1% market share, Amazon pulled the plug and exited China's ecommerce market. Again, the crocodile had beaten the shark.

Jack Ma is a familiar figure globally, and his speeches before the failed Ant Group IPO in 2020 were carefully studied by investors and revered by aspiring entrepreneurs across the world.

However, little is known about why and how he, who in an earlier job search had been rejected by KFC, built one of the world's largest consumer tech companies, or about how the results of the competition in China turned out to be so different from all the rational analyses.

While many of the analyses were focused on Alibaba versus Amazon's tactics and on Ma's personality, we think the real answer lies in the following factors: *people, organization, product*, and, ultimately, *leadership.*

The David-beating-Goliath-type story is not unfamiliar to entrepreneurs in China. Mao Zedong's communist guerrillas beat the better-funded, US-equipped, and more numerous Chinese Nationalists in 1949 to proclaim the People's Republic.

Similar to Chairman Mao's guerrillas, Jack Ma and his 17 cofounders came from humble backgrounds. They did not turn down cushy job offers in order to change the world—rather, they simply did not have other options. In 1999, the year of Alibaba's founding, China's GDP per capita was merely $873—on the same level as Indonesia, and 5% that of the United States.

What's more, an early employee of Alibaba told us that he and his colleagues were working in the office over the weekends because they really did not have anything better to do.

Under Ma's urging, each person adopted a nickname, usually drawn from famous Wuxia (Martial Heroes) novels. Ma's nickname was Feng Qingyang, an elderly sword legend in Hong Kong novelist Jin Yong's *The Smiling, Proud Wanderer*.

The early employees of Alibaba were young and fearless, with nothing to lose and completely without an understanding of work–life balance.

Alibaba also put a lot of effort into its people and organization. Managers at each level were trained extensively for building a team, while chief people officers were appointed at the group, business unit, subsidiary, and department levels. Restructurings have been a regular occurrence to "embrace change," as Jack Ma would put it.

Alibaba's early products were aligned with the key pain points of buyers and sellers: the B2B site Alibaba.com allows thousands of Chinese manufacturers to advertise their products to the world; Taobao's seemingly messy interface allows sellers more space and tools to display their products as trust online was low; Ali Wangwang is the instant messaging software facilitating the communication among all parties such that sellers know individual consumers' need; Alipay was created because banks did not want to provide escrow for small transactions.

And Alibaba did not stop there. When it encounters new challenges, it solves them by building a new solution that will lead to standalone products, companies, or even groups of companies. In addition to various ecommerce platforms, Alibaba Group's ecosystem currently includes a logistics platform (Cainiao Logistics Network), a fintech platform (Ant Group), a tech infrastructure (Alibaba Cloud), local services (Gaode map and ele.me for food delivery, Koubei for food and beverage reviews), lifestyle businesses (Alibaba digital media and entertainment group), and many others (see Figure 1.1).

Throughout all the challenges, Ma was not only the key spokesperson to the outside world, but also the iconic figure and undisputed leader who glued the whole organization together as a coherent force through uncertain, turbulent times.

One might wonder, if one day China's market size is no longer big enough for Alibaba's growth appetite, will the crocodile be able to swim in the open ocean?

YANGTZE CROCODILE GOES INTO OTHER RIVERS

Fast-forward to 2022, Alibaba is now in a position very similar to Amazon's after launching in China many years ago. Expanding globally through multiple subsidiaries and business areas, Alibaba has now become the shark.

It started in 2014, when Alibaba went public on the New York Stock Exchange through the largest initial public offering (IPO) of the year, valuing the company at more than three times the market cap of eBay.

The company, which became the undoubted leader in China's vast ecommerce market, was setting its ambitions globally. In 2015, Ma encouraged his team to set the target to serve at least 2 billion people globally—a goal he has repeated a few times on different occasions.

FIGURE 1.1 Alibaba Group

His successor, Daniel Zhang, the current board chair and CEO of Alibaba, translated Ma's strategic vision into the following execution plan: "To achieve the number (of serving 2 billion customers), we must go global. Our ultimate goal is global buying, global selling, global paying, global shipping, and global travelling."[1]

Since then, Alibaba's cross-border subsidiary AliExpress has helped Chinese sellers reach out to consumers across the world; Alibaba Group has acquired ecommerce platforms including Southeast Asia's Lazada, South Asia's Daraz, and Turkey's Trendyol; Alibaba Cloud has already deployed data centers in Jakarta, Sydney, Dubai, and Frankfurt, among other cities across the globe.

Ant Group, Alibaba's digital financial affiliate, has been even more active globally. Ant Group started from Alipay—a payment and settlement tool used to facilitate transactions on Alibaba's ecommerce platforms. Over the years, it has grown into a digital finance behemoth in China, and its ambitions have not stopped there.

Ant Group is a major shareholder or joint venture partner for Dana in Indonesia, Touch 'n Go Wallet in Malaysia, GCash in the Philippines, True-Money in Thailand, eMonkey in Vietnam, and Wave Money in Myanmar. Ant has also taken control of Lazada's payment subsidiary helloPay.

In addition to Southeast Asia, Ant has become a major shareholder of PayTM, a leading payment/wallet service in India, and similar companies in Korea, Pakistan, and some parts of Europe.

However, Alibaba Group's global expansion has not been an easy journey. It needs to figure out its own strategy and execution for how to adapt its experience, talent, and financial power in foreign rivers and sometimes canals. There is a reason crocodiles did not flourish in every water system.

Just as Amazon and eBay faced in their foray into China, Alibaba now has to deal with the issues and challenges related to its products, its people, its organization, and ultimately a test of its leadership.

But we believe that Alibaba is not Amazon—although they have many similarities and face the same challenges, Alibaba has reacted differently because it is genetically different: its history, culture, and the pathway that led to where it is today are all in contrast to Amazon's.

Alibaba's global journey, therefore, leaves different success factors, and lessons for others, which we will expand and discuss throughout this book, along with experiences and case studies from other major Chinese tech companies.

AMAZON'S PLANNED, CONTROLLED, AND WELL-EXECUTED DEMISE IN CHINA

Amazon shut down its domestic ecommerce marketplace in China in 2019. While there were many analyses that, predictably, blamed "communist China's" protection of local players, others are pointing to more real issues: failure to adapt to Chinese consumers' tastes; refusal to match the offerings of local competitors; the user experience; and overly focusing on trust rather than product variety.

However, this note from an Amazon executive in Beijing in 2015 was probably telling: "According to the forecast our team just completed, Amazon China will expect a steady decrease of sales over the next four years, to its eventual demise."

"So what are you going to do about it?" we asked.

"Nothing, why should I stick my head out and do something about it?" the executive replied. "Decisions are made in Seattle anyway; we just need to go with the cause that is already set in motion."

"Which is?"

"Cut the warehouse and logistic capacity to align with the predicted sales decrease." No wonder Amazon added later on (in 2018), "learn and be curious" to its set of leadership principles, emphasizing that executives should avoid complacency after achieving success. ▨

 THE RIVER THAT BREEDS CROCODILES

So how did Alibaba and its contemporaries—Tencent, Baidu, Meituan, Byte-Dance, and so on—grow so quickly to become giants? To understand this, let's start with a story.

When I was with a group of Chinese investors in Jakarta a few years ago, I remember we were looking out from the lounge of the Westin perched on the 52nd floor. The investors were puzzled: "Why does Jakarta still have so many low-rise neighborhoods? It is inefficient."

The context of their question was the following: in the early 2000s, Chinese cities went through a wave of urban renewal. Existing low-rise neighborhoods were dismantled, giving way to high-rise residential blocks and planned communities. As a result, cities became bland and highly similar to each other, unique characteristics of cities and neighborhoods were lost, and many ordinary folks lost homes where they had been living for decades.

We are not advocating that it was the right thing to do, but the resulting, almost homogenous, cityscape became the bedrock for efficient penetration of ecommerce, food delivery, new retail, and many other business models.

The willingness of the government to dismantle the old for efficiency was, undeniably, paving the way for a national market in which tech entrepreneurship could scale much faster and wider.

Another dismantlement, possibly more profound, was on the societal and cultural front. In the first 30 years of the People's Republic, including through the disastrous upheaval of the Cultural Revolution (1966–1976), China's government dismantled the traditional social hierarchy and powerful, entrenched, local interest groups. Political power, as well as land ownership, was concentrated in the hands of the government.

A long-lasting effect of the upheaval was a clean slate, where infrastructure can be created much faster (e.g., no lengthy land acquisition process). In addition, more than a billion people now speak the same language (standard Mandarin), receive the same education (nine years compulsory primary and secondary), and watch the same entertainment programs.

This might sound trivial, but if you know that China was previously so fragmented that during the World War II, army regiments from different provinces simply could not coordinate with each other because soldiers and officers spoke mutually intelligible tongues (dialects), you start to appreciate what has been done.

The government is also willing to admit its own mistakes and dismantle its own orthodoxy. Throughout much of the twentieth century, successful Chinese entrepreneurs almost all made their fortunes outside mainland China, mainly in Taiwan and Hong Kong, as well as across Southeast Asia. Mainland China, in contrast, experienced years of political, economic, and sometimes military turmoil. There were almost no successful private companies in this period.

A pivotal period in history was the reform and opening up initiated by Supreme Leader Deng Xiaoping in 1978. Over a period of 20 years, pragmatic policies gradually and then entirely took over from orthodox communism. As Deng said, "It doesn't matter whether a cat is black or white; as long as it catches mice, it is a good cat." This shift to pragmatic policies and focus on economic development unleashed the growth potential as well as the entrepreneurial zeal of millions of Chinese.

"Development is of overriding importance," Deng repeatedly said. Successful entrepreneurs, both domestic and foreign, became idols of young people. Jack Ma, Pony Ma (no relation—the founder of Tencent), and many others left their stable jobs in the public and private sectors to create their own companies; they were joined by the likes of Robin Li, who quit his job in the United States to return to China.

On the hardware front, the government has built high-speed rail, expressways, and aviation networks across the country, making the flow of goods and people fast and efficient. This highly reliable and efficient infrastructure is the system that many Indian entrepreneurs and executives are looking for.

Such buildup is almost always planned and executed ahead of time—when Shanghai Pudong International Airport was completed and opened in 1999, many media commentators were still complaining that it was a white elephant—the existing Hongqiao International Airport was not running at full capacity yet. However, barely three years later, in 2002, the new airport was already running at full capacity.

AT FULL SPEED

A critical moment was China's official ascension into the World Trade Organization in 2001, after more than a decade of negotiations.

Since then, China's economy, already unshackled from central planning by Deng's reforms, started growing in leaps and bounds. With supportive government policies, fast-improving infrastructure, competitive wages for an educated workforce, and newfound access to global markets, China's manufacturing sector took off.

Foreign direct investment (FDI) inflow to China grew in the past two decades from $40 billion in 2000 to $163 billion in 2020,[2] surpassing that to the United States. Over these two decades, China has managed to build a large, deep, and sophisticated supply chain where the benefits of integration outweigh the labor cost increases. The much-talked-about shifting of supply chain out of China at the beginning of the pandemic did not happen. ▨

State-owned telecommunications companies established almost-nationwide 4G (and now 5G) mobile networks, even in the lifts of residential areas, as well as remote, sparsely populated, and less-developed areas in western China.

Tapping into this infrastructure, Chinese consumer electronics manufacturers quickly seized the opportunity. According to government statistics, in

2010 fewer than 300 million people were connected to the internet; by 2015 that number had surpassed 600 million; in 2020, more than a billion people were connected.

In addition to people and goods, now information and data can flow freely, and direct to hundreds of millions of consumers, as well.

EVERYONE IS KYC-ED

Every Chinese person opening a bank account had to be KYC-ed (know-your-customer, to verify you are who you say you are) physically by the bank. You need to present your ID card (with a chip embedded), verify your biometrics, and match your mobile number.

This might sound trivial for tech, but that is precisely what allowed China's mobile payment providers Alipay and WeChat Pay to onboard hundreds of millions of new users without having to go through the costly KYC process themselves. All they need to do is to ask the customers to bundle a bank account and verify their mobile number.

This little-known fact was actually one of the key enablers of the mobile payment's rapid ascent. ■

In effect, China's government has created, over a relatively short period of time, a gigantic, homogeneous market with a large population and extensive consumption power. The United States is the only other country in the world that fits these criteria and matches the size.

It is interesting to note that the worries that the Chinese people had in the 2000s about traditions being lost turned out to be unfounded. When the economy developed, many of the traditions feared lost began coming back even stronger. For example, people, including the young, nowadays attach more importance to traditional festivals compared to 10 years ago. Han-style clothing (the traditional dress of the Han Chinese people) has become more popular with Generation Z.

This infrastructure leap also shaped the view of internet practitioners in China, in the same way the great firewall of China (an artificial barrier erected on the internet to keep international giants such as Facebook and Google outside) has. They are in an ecosystem that is unique, and those who do not appreciate this fact when they venture out of China typically hit obstacles very soon.

POLICY PUSH FOR ENTREPRENEURSHIP SINCE 2014

In capital markets, there is a saying: "When the Fed sneezes, the whole world catches the cold." Every word of Jerome Powell, the current chairman of the Federal Reserve, is watched and analyzed carefully for clues and hidden messages, and billions of dollars can be moved as a result.

In China, billions of dollars can be moved as well when someone from the top leadership signals a policy focus or shift.

Premier Li Keqiang's promotion of *mass entrepreneurship and mass innovation* (大众创业、万众创新) in 2014 spurred a great wave of tech entrepreneurship activities.

MASS ENTREPRENEURSHIP AND MASS INNOVATION

Premier Li first introduced the notion of "mass entrepreneurship" in a speech at Summer Davos held in the Chinese city of Tianjin in September 2014. In the speech he said:

> Leverage the current of the reform . . . to create a new wave of "mass entrepreneurship" and "grassroot entrepreneurship" across the 9.6 million square kilometers territory of China, to form a new state of "mass innovation," "everyone's innovation."

He reiterated that the internet is the key driver of such entrepreneurship and innovation in another forum two months later. During the National Congress plenary in March 2015, Li further stressed the need to make mass entrepreneurship and mass innovation "the twin engines" of continuous economic development of China. He mentioned "innovation" or "entrepreneurship" 63 times in the government report speech that year.

The lingo of Chinese officials can seem awkward to external audiences who are not used to its political context. But in this instance the message is very clear. ▪

Just as the Fed chair does, Chinese leaders deliberate policy speeches very carefully before they come out. Once the words are out, people across the country interpret them, and act on them.

Billions of dollars were set aside to invest in venture capital, and tens of thousands of startups were created. While the biggest internet companies were probably not a direct result of this, and wastages were created during the process, the movement indeed created whole suites of startups in various sectors, making the whole tech ecosystem much deeper and richer.

Premier Li did not stop there. A year later, he put forward another slogan, "Internet plus" (or "Internet +"), which calls on entrepreneurs to use the internet to enable and empower traditional industries such as financial services, manufacturing, logistics, education, and healthcare. One by one, industries are disrupted, sometimes in a messy way. We believe Premier Li understood that many of China's traditional industries were not as sophisticated or competitive as their Western peers—and the internet could be a way for them to achieve sector-wide efficiency gains, and leapfrog ahead.

Different from the perception that the government directly funds or controls the process, the role of government is to use policy as a lever to set the direction to encourage entrepreneurs and create an environment in favor of exploration and risk-taking.

THE GREAT FIREWALL HURT, NOT PROTECTED, CHINESE INTERNET COMPANIES

Many have also attributed Chinese tech giants' success to the great firewall of China, an artificial barrier erected on the internet to keep international giants such as Facebook and Google outside.

The great firewall did not create a protected environment for certain champions; instead, it encircled an ecosystem of fierce competition. Even founders of the most successful internet companies cannot rest on their laurels but must constantly fend off existing and upcoming competitors. Alibaba is now challenged by Pinduoduo; Tencent is trying hard to defend against ByteDance; Baidu, whose leading position in search and online advertising once seemed solidified, has now fallen far behind.

Without the existence of the wall, the same scenario would have probably emerged anyway. Western tech companies that were allowed to compete fairly in China, including Amazon and eBay, Groupon, and Uber, lost out to domestic players Alibaba, Meituan, and Didi, respectively.

The wall might be a drawback even for Chinese tech giants. Although they have developed globally competitive products, capital, talent, and organizational capabilities, the fairly enclosed ecosystem actually becomes an impediment when these companies venture out, as all the competitive advantages they have are quite often heavily customized to the Chinese market.

American internet companies, in contrast, have a global audience from the onset, and a global English-speaking talent base to tap into. ■

NOT MAGIC, BUT RIDING THE WAVES OF LUCK AND TIMING

As you can see, the success of Alibaba is more than just pure grit and talent alone. Of course, there are tactical plays that Alibaba made—and this is a testament to its cunning, strategic agility, and ability to fight wars to create and develop market share. We will cover these in the next few chapters. However, there is a lot more under the surface.

To succeed in China in the early part of this century, you had to understand the local market. You had to be aware of what was being built and how your business could benefit from (or be killed by) it. You had to be in the right place, at the right time, with the right people, and working harder than your competitors.

These lessons are still applicable today. These are the macro areas you need to look into when you try to understand Chinese players, either as a potential partner or a competitor.

While the failures of eBay and Amazon in China were mainly due to their mistakes—many of their tactical initiatives suggest that they failed to understand the Chinese local market and its differences from that of the United States, as many pedagogical materials and research reports described—we cannot deny that the Alibaba miracle relied on several key enablers of the macro environment in China, which include:

- Being the world's factory provides surplus products for consumption (from the supply side).
- Decades of economic growth led to a large number of middle-class families with stable income (from the demand side).
- Improved infrastructure reduces transaction friction and helps to build a large domestic market.
- Smartphones and mobile internet that connect hundreds of millions of consumers online over a short period of time.
- A large workforce with good education and a diligent working attitude.
- Stable government and pro-economic development policies.
- A highly competitive domestic market that cultivates a generation of talented employees and entrepreneurs.
- Capital from overseas and subsequently successful entrepreneurs.
- Fast growth and successes of tech companies inspire many other talents' youth to become tech entrepreneurs.

Overall, it is not magic or a simple centralized government maneuver, but a confluence of various factors in a particular timing. Among investor friends we often debate whether the same would happen in other large emerging countries. Our conclusion is that yes, tech will evolve, but in a different way compared to how it has in China.

NOTES

1. Alizila Staff, "Alibaba CEO Zhang Tells Employees: 'We Must Absolutely Globalize,'" News from Alibaba, May 14, 2015, https://www.alizila.com/alibaba-ceo-zhang-tells-employees-we-must-absolutely-globalize/.
2. United Nations Conference on Trade and Development, *World Investment Report 2020: International Production Beyond the Pandemic* (Geneva: United Nations, 2020), https://unctad.org/system/files/official-document/wir2020_en.pdf.

CHAPTER TWO

Strategy and Tactics of Chinese Tech Companies

Strategically we must despise all our enemies, but tactically we must take all of them very seriously.

—Mao Zedong, leader of the communist revolution, 1957

WITH ALL THE TAILWINDS and headwinds described in Chapter 1, you might think that the Chinese tech companies that became successful were incredibly lucky. The truth is, yes, they were lucky to have been at the right place at the right time, but their success was in no way guaranteed.

Because for each of the sectors and business models, you would have dozens, if not hundreds or even, in some cases, thousands, of companies competing for the same prize. Success, importantly, also depends on the entrepreneurs, their strategies, and how they execute those strategies.

In this chapter, we look at how many entrepreneurs are drawing from the thoughts of Mao Zedong, who led China's communist revolution, how copying becomes a key driver of innovation, and how to get people together to not only work effectively toward a dream, but also embrace constant changes and endure hardship.

These are the critical factors that you would see in each of the successes in China's tech sector.

 ## CHAIRMAN MAO'S STARTUP PLAYBOOK

Interestingly, many successful or struggling entrepreneurs we personally know have been reading the five-volume *Selected Works of Mao Zedong* in recent years and drawing inspiration from the leader of the communist revolution. This is quite interesting because they are probably the biggest beneficiaries after Mao's societal and economic policies were lifted by Den Xiaoping. For them, Mao's writing is a contemporary of the *Art of War*, an ancient text written by Chinese military strategist Sun Tzu more than 2,000 years ago.

In fact, some entrepreneurs half-seriously said that the Communist Party of China, regardless of whether you agree with their ideology or not, is a great startup success.

Mao Zedong's military and political strategy led to his victory against the Japanese invasion, against the much better equipped Chinese Nationalist forces in the civil war in the late 1940s, and again in achieving a stalemate against American forces in Korea in the 1950s.

How did they win? With a playbook that recognizes when to use 游击战 (guerrilla warfare) and when to use 运动战 (mobile warfare). Guerrilla warfare is self-explanatory, whereas mobile warfare keeps moving the military units, and always making sure of military superiority in each specific battle, in order to wipe out entire enemy units, one after another.

Depending on the circumstances, military commanders will choose whether to use guerrilla warfare versus mobile warfare. When the enemy is strong, the right strategy should be concealing the forces as small, mobile, and independent guerrilla units. A 16-character mantra was distributed among the commanders of various units: "敌进我退, 敌驻我扰, 敌疲我打, 敌退我追"—"We retreat when the enemy is advancing; we harass when the enemy is camped; we strike when the enemy is tired; and we chase when the enemy is in retreat."

Once sufficient military strength is built up against the enemy, the strategy becomes mobile warfare. The Chinese government adopted exactly the same methodology in the Korean War—although the Chinese forces were weaker than their US foes by a large margin, at each specific battlefield they were superior in both numbers and vantage points.

Another example of Mao's strategic thinking is the notion of using the countryside as a base to encircle and finally capture cities (农村包围城市).

The conventional wisdom and doctrine in the early days of the communist revolution was to follow the Russian experience: organize the workers in cities to protest and eventually seize power. The strategy, however, failed to work in China the way it had in Russia.

Mao, an avid reader of Chinese history and a former librarian, challenged the notion. He believed that China was still a largely subsistence farming society, and the real power was in the countryside. To succeed in the revolution, the communists must abandon its worker-focused doctrine to set up bases in the countryside, accumulate power, and eventually surround and take the cities.

He did not say that workers in cities were not important. In fact, he always believed that workers should be the core of the revolution together with the peasants. However, his deep sense of reality made him focus on the countryside in the beginning.

When the communists eventually accumulated enough power through their countryside and mountain bases, their attacks on the cities were both swift and decisive.

Following this methodology, Alibaba built its ecommerce empire through first addressing the customer-to-customer (C2C) market, focusing on low-price products and individual sellers. Once the sales volume became big enough, and the scale brought efficiency, Alibaba moved up the value chain to build the brand-focused Tmall.

Pinduoduo has been using the same strategy to challenge Alibaba in ecommerce, while Meituan built its online travel business precisely starting with the masses.

This is in contrast to the conventional wisdom that ecommerce companies should focus on highly transactional sizes and good margins. Uber started with a black car service before moving to the mass taxi market, while Tesla's first model was a premium vehicle, before using that to prove its name to scale the mass production.

There is no right or wrong here, just which strategy fits the current stages of the market more. In this, Mao has said that 没有调查就没有发言权 ("Those who have not conducted investigation shall have no right to speak"). One of Mao's essays studied vigorously by entrepreneurs is the 1930 Report from Xunwu ("寻乌调查"). In this more than 80,000-word-long text, Mao detailed the history, agriculture, business, artisanry, land ownership, and major social conflicts of Xunwu, a small, remote, and mountainous county in Jiangxi province. The detailed assessment helped Mao form the revolutionary strategy.

A few entrepreneurs we know use this report to benchmark their own assessment of the market they are in, the strategies they will take, and the products they will develop.

COPYING, EXTREME COMPETITION, AND BEYOND

> Let you start first, and I will go for the kill with the model you have explored, in a market you have educated. In your eyes, I am your competitor; in my eyes, you are my tool.
> —Duan Yongping, Chinese entrepreneur and billionaire investor

Another tactic "virtue," controversially, is copying. Entrepreneurs ruthlessly copied tech products and business models not only from the United States, but also from each other.

As a result, any emerging business model will quickly descend into an aggressive and messy battle of dozens, hundreds, and, in some cases, thousands of companies, often doing exactly the same thing.

Unlike in traditional industries where it would take months if not years to build up the supply chain, distribution network, and customer base, with tech and mobile internet infrastructure, firms' every timeline is compressed.

This is something that many foreign companies operating in China then (but maybe less so now) didn't get.

In 2010, Groupon, armed with its success in the United States and Europe, decided to venture into China with its business model of group buying.

Almost overnight, thousands of group buying sites emerged in China— including Manzuo, Lashou, and Meituan. At one time, more than 5,000 group buying businesses were registered in China.

In what was dubbed the famous "war of a thousand group-buy companies" (千团大战), the belligerent companies employed almost exactly the same business model as Groupon and competed fiercely with each other. Some players employed ground sales teams whose sole purpose was to sabotage competitors by offering merchants better terms to switch over.

Groupon's European managers were unaware of the competitive environment they were getting themselves into. They followed a successful playbook, but they were just unaware of the aggressiveness of the competition and were not able to react nimbly enough. Groupon's first merchant was hijacked by Lashou's ground team before the deal went live. That is still a folktale told by the entrepreneurs in China.

Similar stories were repeated many times in the decade that followed: ride hailing, O2O (online to offline), shared bicycles, shared umbrellas, power banks, live-streaming, cross-border ecommerce, community group buying, and so on. Even the more technologically sophisticated areas of smartphone manufacturing, artificial intelligence, face and voice recognition, virtual reality, and electric cars were not spared.

By now you should know that the business situation is not about just copying—because Groupon was copying its successful playbook into China as well. It is actually about many players ruthlessly copying from each other, at an intensity not seen in other markets.

In a closed ecosystem, protected by the Chinese firewall for good or for bad, internet firms and business models in China have their own trajectory of development. The gigantic market with fierce competition has the potential to breed unique species.

Investors and entrepreneurs have always known that only a small number of players could survive, and then thrive. During the process, the market is educated, experienced talent is groomed, and fortunes are made (for some).

Such competition also embodies a system of fast innovation through competition and rapid iteration. Players have to make incremental innovations to be competitive in order to survive, and when hundreds of players do that, the whole sector evolves rapidly.

For example, when smartphone manufacturers in China were competing fiercely with little differentiation, some started to innovate: construction workers needed a mobile phone with seven stereo speakers; African traveling salespeople needed a mobile phone that can last a month without charging. Incremental innovations were designed, rolled out, and quickly tested—in a fashion very similar to the concept of design thinking that is being taught in business schools across the world.

JUST COPYING IS NOT ENOUGH

Former and current executives of Huawei we spoke to point out that the key success factor is relentless willingness to do the extra to make customers happy.

"Initially, Huawei and ZTE actually recognized that their products were inferior, but to survive they decided to compensate for that disadvantage with better services and fast problem solving," one former Huawei executive told us. "Large Western companies typically face very different competitive environments and product advantages, therefore it is difficult to develop the internal urgency to serve across the organization."

Even to this day, Huawei still embodies the spirit to serve the customers to the extreme. The CTO of a large consumer internet company told us that whenever they have a problem with their network, even though it might not be caused by Huawei products but by those of Huawei competitors, Huawei's service engineers will quickly jump onsite to try to fix it anyway, with no questions asked about whose responsibility it really should be. ■

In summary, individual battles might be concluded very quickly, with winners and losers decided and lessons learned ascribed to the history book of Chinese tech.

Entrepreneurs who fail often start building other businesses—keeping the whole competition engine running. However, winners of such battles can hardly sit on their laurels and enjoy the spoils. New competitors keep emerging, often from incumbents encroaching from adjacent industries to participants of emerging disrupting models.

In contrast, many American entrepreneurs were born in privileged backgrounds with early access to computers. They often have a greater vision and determination to innovate and change the world. American society also prizes original, revolutionary ideas—to the extent that Mark Zuckerberg, founder and CEO of Facebook, is still reprimanded in the ecosystem for copying other people's ideas.

GET PEOPLE TO WORK TOGETHER

By now you have a good sense of the environment Chinese tech companies grew in, as well as the strategies and tactics they have used. However, no matter how lucky they have been or how sound their strategy is, at the end of the day it is people that make a difference.

In a constantly intense competitive environment, how do you get your people to work hard and beat the competitors? How do you organize them in the most effective way possible? People and organization occupy the mind of founders of large tech companies, and are among the most discussed topics among entrepreneurs.

Alibaba's Jack Ma built himself up as an icon to inspire people, but that was not enough. His cofounder, Lucy Peng, devised a whole set of values, methodologies, and organizational practices to ensure that people across the various levels of the organization are aligned. We discuss this concept in more detail in Chapters 5 (People) and 6 (Organization).

An obvious sign of these tech firms' abilities to get their people to work hard is that Jack Ma of Alibaba and Richard Liu, founder of JD.com, both publicly advocated "996 culture"—working from 9 a.m. to 9 p.m., six days a week—and claimed that 996 culture is a blessing. One of Alibaba Group's six core values is "If not now, when? If not me, who?"

IF NOT NOW, WHEN? IF NOT ME, WHO?

(One of the Six Core Values of Alibaba Group)

// This was a tagline in Alibaba's first job advertisement and became our first proverb. It is not a question, but a call of duty. This proverb symbolizes the sense of ownership that each Aliren [Ali people] must possess." ▪

Large tech companies asked their employees to put in extra effort and time to achieve their professional success. In return, employees are rewarded with not only extra pay, but stock options which they hope one day will bring them financial independence. In the early 2010s, large internet companies overtook multinational companies (MNCs) and financial institutions as the preferred employers of fresh graduates.

And as mentioned in Chapter 1, early employees of large tech companies often had little else to do in their social life, and chose to dedicate their time to work instead.

For many of them, the purpose was not to change the world, but to create personal wealth, and bring pride to their families.

CHANGE IS THE ONLY CONSTANT

Aside from our dream, the only thing that does not change is change itself.
—Jack Ma

The fierce competition Chinese tech companies are subjecting themselves to is very hard to really appreciate unless you are in that environment. The fact that Groupon, Uber, and Amazon, known to the world as among the fiercest and most aggressive tech companies, all lost in their respective ventures into China probably provides a sense of the challenges that foreign companies face in China.

In addition to product and operations, many of the Chinese tech giants also iterate their organization at a speed never seen before (for large, complex companies). A malleable organization becomes essential when constant strategic adjustments are needed to adapt to the rapidly changing external environment. When the market is full of uncertainty and the future path is ambiguous, organizations need to be ready to change and adapt.

However, change is against our human nature. Mentally we may understand the need for change, but psychologically we prefer the status quo because changing is usually accompanied by uncertainty and anxiety. That is the reason "leading change" is a popular executive program offered by business schools.

Jack Ma of Alibaba Group has been a key champion of embracing change. He professed that gospel in multiple speeches to internal and external audiences.

In 2011, Alibaba's executives had a heated debate on the company's future strategic direction—which model to pursue for Taobao, its main consumer ecommerce platform. Instead of running more strategic exercises, Ma decided to split Taobao into three entities: eTao, Taobao.com, and Taobao Mall (Tmall), and let them compete against each other.

It is not easy to restructure a well-established company and reassign its people to competitive positions. But Alibaba achieved this in a very short period of time.

Indeed, Alibaba conducts small-scale reorganizations almost continually throughout the year. We know personally someone who has had five different bosses over 1.5 years—a record—and another person we know went through six reorganizations over her three-year stint at Alibaba.

Similarly, Huawei forces their executives to switch posts and countries once every two to three years. They intend to create a culture that continuously changes and avoids organizational stagnation—a common problem facing large, successful MNCs.

CHANGE IS THE ONLY CONSTANT

(One of Six Core Values of Alibaba Group)

II Whether you change or not, the world is changing, our customers are changing, and the competitive landscape is changing. We must face change with respect and humility. Otherwise, we will fail to see it, fail to respect it, fail to understand it, and fail to catch up with it. Whether you change yourself or create change, both are the best kinds of change. Embracing change is the most unique part of our DNA." ▪

This nimbleness and constant change allow large Chinese tech companies to adapt to the fast-changing environments in their home market. Senior leadership, who senses the shifts in the market, could quickly deploy people and resources to address changes, fix problems, and seize opportunities. This also keeps executives on their toes and prevents slack.

However, this methodology and mentality proved problematic in Chinese companies' overseas ventures, which is discussed in more detail in Part Two.

HARDSHIP IS A VIRTUE (FOR FOUNDERS)

Mao achieved his success in revolution not only through his brilliant strategies and tactics. Throughout the revolution, he endured tremendous hardship: his first wife was executed by his enemies, he lost two of his three sons, and he was sidelined by Russian-educated and more cosmopolitan factions of the communist leadership a number of times.

Many successful Chinese tech entrepreneurs have also faced enormous hardships. First-generation Chinese tech entrepreneurs are also reminded that the endurance of hardship is a virtue that will contribute to future success.

Huawei has repeatedly used the hardships and restrictions it has suffered, including the bans by the Trump administration, as tools to motivate its workforce—its founder, Ren Zhengfei, endured significant hardship during his early years, through the cultural revolution, military career, and a few near-death experiences while turning Huawei from a small electronics manufacturer to a global network and consumer electronics giant.

In a similar fashion, Alibaba has included "must have suffered hardship before" as a recruitment criterion in hiring key managers—a reflection of Jack Ma's early days where he was rejected in, among many occasions, an interview for a part-time job at KFC. One of the cofounders of Alibaba told us that when he was joining Ma in the mid-1990s, his mother had serious reservations because Ma "looked like a swindler."

The notion about hardship is actually about resilience and perseverance because managers know that the undertaking will be hard. In this way, Ben Horowitz, a Silicon Valley entrepreneur and partner of venture capital firm Andreessen Horowitz, summed it up in his book *The Hard Thing about Hard Things*:

> Whenever I meet a successful CEO, I ask them how they did it. Mediocre CEOs point to their brilliant strategic moves or their intuitive business sense or a variety of other self-congratulatory explanations. The great CEOs tend to be remarkably consistent in their answers. They all say, "I didn't quit."

CHAPTER THREE

Everyone Is Going Global

The companies that have been hardened in the Chinese market, even those who are not that outstanding, should now enter and practice themselves in the global arena. They already have the skills to go global.

—Luo Zhenyu, Chinese online talk show host, 2017

FROM THE MID-2010S, CHINESE entrepreneurs and companies, buoyed by the experiences, capital, and talent accumulated through rapid advances in domestic markets, started exploring global markets. One theory behind this global expansion is "time machine management," which was described in SoftBank's annual report of 2020.

Masayoshi Son, the bold founder of Japan's SoftBank, is now a household name. His Vision Fund, with more than half its money coming from Saudi Arabia and United Arab Emirates, had made a splash in the late 2010s, backing high-profile companies including WeWork, Uber, DoorDash, and Opendoor.

To the Chinese, Son is a familiar figure as an early backer of Alibaba Group, which still counts SoftBank as its largest shareholder. Oddly, most Chinese tech entrepreneurs, executives, and investors are very familiar with Son's theory of time machine management, which seems little known outside China.

The theory was actually not a fable, but was discussed extensively in Soft-Bank's annual report. In the extensive document, time machine management is described as "fostering the global incubation of superior internet business models developed in the United States."

The rationale behind this approach is simple: new business models emerged in the United States, where economic development, infrastructure, and tech penetration was superior. Sooner or later, the same business models will emerge in other parts of the world, and SoftBank should be an active participant and investor in this process.

A decade later, Germany-based Rocket Internet implemented this strategy (without mentioning the name) thoroughly and launched hundreds of internet businesses across Europe, Latin America, Asia-Pacific, the Middle East, and Africa.

Are Chinese tech firms and their business models going to be replicated and succeed in other regions, just like the time machine theory predicts?

GUERRILLA FORCES: SUDDENLY THEY ARE EVERYWHERE

In December 2016, one of us was in Dubai. The air was full of excitement that the ecommerce sector was on the verge of taking off. A month earlier, a group of investors led by the chairman of Dubai's largest property developer, Emaar Group, and Saudi Arabia's Public Investment Fund announced they would invest US$1 billion to create a new ecommerce platform called noon.com; later, in the same month, Bloomberg reported that Amazon was in talks to acquire Souq, the region's largest ecommerce site, which had become a unicorn (a privately held startup company valued at over US$1 billion) in a US$275 million round earlier that year.

Amid the excitement, a few executives I spoke with had some slight concerns. "There is this company that came from nowhere, is growing really fast in the Saudi market, and looks set to overtake Souq there soon," said one logistics executive who had noticed JollyChic.

JollyChic is based in Hangzhou, a city that was little known in the business world before hosting Alibaba's headquarters. You could hardly find any information online, aside from its product promotions and app install ads—no media interviews with the founder, no case studies about its practices, and even no LinkedIn profile for its key executives. This is what we call *guerrilla forces*, which means that it is hard for external people to find information about it in the public domain.

The company, founded by veteran ecommerce executive Aaron Li, by then already had a headcount of close to 2,000, mostly in China but also including a large customer service team based in Jordan. Yet almost nobody we spoke to in the Middle East even knew who the founder was, which city the company was based in, or how big it had become. People struggled to understand why this unknown company could catch up so fast, unsettling Souq, which had been operating in the region since 2005.

Three months later, in March 2017, Amazon concluded the deal to acquire Souq, at a valuation of US$580 million, or a 42% discount from the previous valuation. Competition from not only Noon but also JollyChic was an important factor in the urgency to sell.

That same month, I was in Jakarta, Indonesia. Again, the chatter was about a little-known company almost suddenly appearing everywhere. J&T Express, by then a two-year-old ecommerce logistics company, already had more than 10,000 employees in the country. A few executives from rival ecommerce logistic companies I spoke with were struggling to understand how J&T could grow so fast. "Our experience shows it takes years to build logistics operations in the challenging landscape of Indonesia," one said. Again, J&T had almost the same level of secrecy as JollyChic: no media interviews, no case studies, no LinkedIn profiles.

Similarly, when SheIn (usually seen as Shein in the United States)—a cross-border ecommerce player—overtook Amazon as the most downloaded shopping app on US iOS and Android app stores in May 2021, most people over

WHO IS J&T? ECOMMERCE LOGISTICS LEADER FROM SOUTHEAST ASIA TO CHINA

J&T, founded by former Oppo distributor Jet Lee (no relation to the famous actor), became a leading ecommerce logistics player in Indonesia and subsequently across Southeast Asia. Not only that, after careful preparations, in 2019 it launched operations in China, which was famous for cutthroat competition. By the third quarter of 2021, it was delivering more than 20 million parcels in China per day, with a market share of close to 20%. ▨

the age of 30 had not heard of it. One reason is that SheIn only exists in the virtual world, as a pure online digital player, which is different from Zara and H&M, which operate many brick-and-mortar stores. The other reason is that its leadership team had not given any interviews to public media. However, SheIn has achieved remarkable market penetration in the United States with a focus specifically on female Gen Z consumers. Its growth has been further fueled by the global pandemic when consumers relied more on online shopping.

A Google Trend keyword search comparison between SheIn and Zara in the United States shows that the search interest in SheIn got close to Zara's in 2019 and then beat Zara in January 2020 (Figure 3.1). SheIn also surpassed Zara and H&M in US fast-fashion sales.[1]

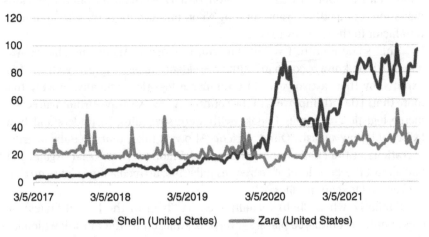

FIGURE 3.1 Google Trends keyword search: SheIn versus Zara

These episodes have something in common: a Chinese or Chinese-founded tech company suddenly grew in leaps and bounds in a large foreign market, leaving local executives in rival companies startled about how these new entrants could amass the expertise and resources to grow so fast.

Unlike the flamboyant Jack Ma, founders of these companies usually keep a low profile—you can hardly find any press releases, media quotes, or interviews in the public domain. A tech journalist with a major global publication who has been following a number of such companies once told us his frustration: "Over four years of talking to their PR teams, I have never managed to extract a single meaningful quote from any of them."

While the world was surprised, this is actually a classic strategy from Mao's guerrilla warfare playbook. Beyond what we see on the surface, there are deeper factors of POP-Leadership that led to successes and failures, and what might happen in the future. We explore more about this in Part Two.

Interestingly, several executives we interviewed told us that the confidence of Chinese companies going overseas was actually partially driven by the strong competition at home. Executives often believe that if they could survive China's hypercompetitive domestic tech market, they would not be afraid of any international competition.

SECOND WAVE: APP FACTORY FOR THE WORLD

Aside from the guerrilla forces and high-profile Alibaba, other big Chinese tech companies were also going overseas. They took a very different approach.

In 2017, ByteDance, known to Chinese as the "app factory," bought Music .ly, a lip-syncing video sharing app with 240 million registered users mainly outside China. The acquisition was a win-win situation—Music.ly had a good head start in international markets but found it hard to grow even further because of lack of capabilities, especially in further product development, content generation, and video recommendation. These are exactly the strengths of ByteDance.

Over the next three years, TikTok, as Douyin's international edition, grew in leaps and bounds.

The founder of ByteDance, Zhang Yiming, wanted to capitalize its core strengths and capabilities to create not only news and entertainment, but also online education, workplace collaboration, ecommerce, and gaming, across the world.

At the beginning of 2020, Zhang delegated the management of the domestic Chinese business to his lieutenants and made a pledge to travel more to make ByteDance a truly global company.

The pledge was ill-fated due to the COVID-19 pandemic that very quickly engulfed the whole world. However, ByteDance's expansion carried on. In that year, it expanded the headcount of its international headquarters in Singapore by almost tenfold. In 2021, ByteDance started experimenting with ecommerce on TikTok in Indonesia and the United Kingdom, off the back of its ecommerce success in China.

I hope everyone will not pay attention to short term glory or loss, instead focus on patiently doing the right thing well. It is about a big vision, but a small ego.

—Zhang Yiming, Founder of ByteDance, 2020

ByteDance has also acquired Moonton, developer of *Mobile Legends*, the most popular multiplayer online battle arena (MOBA) game across many countries in Southeast Asia.

For the first time, one large Chinese tech company is in the position of seriously challenging a US incumbent (Facebook) in its home market, and it is also taking over the world.

Alibaba, Tencent, and Baidu (referred to as BAT) formed the first wave of Chinese companies that had success gaining market share in China and had started to expand overseas; ByteDance, J&T, JollyChic, and SheIn are examples of the second wave.

Most people recognize these companies now that they've popped up, acquired a company, and are on their way to global expansion (or global demise).

However, this is just the tip of the iceberg. How did JollyChic enter the Middle East market with no one noticing; how did J&T grow so fast in two years to be the leader in logistics in China; and how did ByteDance conquer the US market whereas Musc.ly and Kuaishou (another Chinese short-video sharing app) couldn't? In a highly competitive market of fast fashion, with dominant players such as Zara and H&M, how did SheIn find a way to sneak in and then challenge the incumbents?

We believe that the secret lies in the culture of POP-Leadership—and in Part Two, we examine this in detail, and what we can learn from these companies about what to do (and also what not to do).

NOTE

1. Tatiana Walk-Morris, "SheIn Surpasses H&M, Zara in US Fast Fashion Sales," *Retail Dive*, July 12, 2021, https://www.retaildive.com/news/shein-surpasses-hm-zara-in-us-fast-fashion-sales/603160/.

PART TWO

POP-Leadership

CHAPTER FOUR

Leadership

First, they all are extremely focused and market driven. . . . Second, they all are practitioners of long-termism, aggressive and ambitious. . . . Third, they all respect business rules, but yet have reshaped business rules in many industries. . . . Finally, they all have a global vision.

—Neil Shen, founding and managing partner of Sequoia China, on the founders of Meituan, ByteDance, and Pinduoduo

MANY OF YOU MIGHT already know the Monkey King, the legendary Chinese mythical figure who possessed supernatural powers and a rebellious spirit. Wreaking havoc in heaven and hell, as well as the Dragon King's palaces in the oceans, the Monkey King, with his free spirit, was the idol of many young people in China.

However, people tend to forget that the Monkey King, with his superpowers and personality flaws, was actually a disciple of Tang Sanzang, a Buddhist monk (based on a real historical figure), in the sixteenth-century mythical novel *Journey to the West*.

In the novel, Tang Sanzang embarked on an epic journey to India to obtain sacred Buddhist texts. The decade-long journey involved 81 hardships (traditional Chinese texts liked to use multiples of 9), including harsh terrains, bandits, armed conflicts, seductions, and all kinds of demons.

THE TEAM IN *JOURNEY TO THE WEST*

Tang Sanzang could not complete the journey alone. With the Buddha's help, he assembled a team of the condemned: the Monkey King, who was punished for wreaking havoc in heaven, as well as the Pigsy, the Sandy, and the Dragon Prince, who were sanctioned for various violations of heavenly laws.

While all the team members are loyal and determined, they each have their own personalities, often flaws. The Monkey King was powerful, but very impatient and sometimes lacking self-control. He often broke rules, causing trouble here and there. The Pigsy was a bit lazy, sometimes selfish, timid, and horny. Sandy was honest and grounded, if quiet and unimpressive. The Dragon Prince was polite, composed but could unleash power in times of stress.

In a modern team context, you can think of the Monkey King as the star performer who sometimes disrespects rules; Pigsy as the person who brings fun and gossip to the team; and Sandy and Dragon Prince as the majority of the ordinary team members, who can withstand hardship, just want to get the job done, and do not need any shine. ▪

Jack Ma once said that Tang Sanzang was probably the best entrepreneurial leader in (fictional) history. Sanzang did not have any superpowers as his team members did, or any power at all. However, he had the determination to not give up until the ultimate goal was achieved: hardships, challenges, seductions—nothing changed his resolve.

He also pulled various strings to rally the team together and keep them motivated throughout their journeys. He would not hesitate to punish the Monkey King if the latter broke rules. At the same time, he also understood the team members' emotions and often talked extensively with them and convinced them of the purpose of their journey.

Everything we talk about in Chapters 5, 6, and 7 (Product, Organization, and People, respectively) starts with leadership. Who is ultimately in charge, what is their style and character, how do they rally and motivate their team, and how do they resolve problems and overcome challenges?—these are critical to the survival and success of the company they have founded, as well as the confidence of employees, partners, and customers.

To understand Chinese tech companies' overseas journey, commitment, performance, and even their problems and challengers, we need to zoom in on the most important element: leadership.

OUTSIZED IMPORTANCE OF THE FOUNDER CEO

As most tech companies in China are still relatively young, the founders are often still at the helm, and they are pivotal to the companies' continuous success (or failure). In fact, during our interviews of 30+ executives for this book, a key refrain is the critical role of the founder CEO. Without their leadership, companies will find it difficult to navigate the competition in China. A few investors who have been following Chinese tech stocks expressed similar sentiments to us.

To give an example, in March 2021 Colin Huang decided to step down as chairman of Pinduoduo, the ecommerce giant he had founded. This spooked investors, and Pinduoduo's share price dropped almost 23% in a week following the resignation, wiping out almost $50 billion of market value.

The feeling most commonly shared with us at that time was disbelief that Pinduoduo could still supercharge its growth without Colin Huang at the helm. "Essentially you are flying a plane while still building it, do you really want to afford a change of pilot at this time?" a veteran tech investor shared.

Broadly speaking, it also indicated investors' unease with a hitherto-successful Chinese tech company when the founder has departed. In other words, the founder CEO effect is stronger in these Chinese firms, for better or worse.

In comparison, during the week after Jeff Bezos announced he was stepping down from Amazon in May 2021, Amazon's stock price barely moved 1%. This was even considering that Bezos was planning the highly risky undertaking of going to space through a rocket designed by Blue Origin, another company he had founded.

In addition to Amazon, many American tech giants, including Microsoft, Apple, and Google, have already experienced at least one round of leadership succession. While in China, among the first two generations of internet giant founders, only Jack Ma has formally retired.

CHARACTERISTICS OF CHINESE LEADERS: LEARNING FROM HISTORY

We mentioned *Journey to the West*, which was fiction. However, in addition to that book, almost all successful tech leaders we know of are avid readers of actual Chinese history. Why?

If *Journey to the West* is an inspiration to us on this journey, history gives some clues on how and what to expect in this uncharted territory.

China has a deep, complex, but, more important, perfectly recorded history over thousands of years. China's first recorded dynasty, which started more than 4,000 years ago, already had official historians in the court; and each new dynasty's first task after founding was to compile an official history of the previous dynasty.

Chairman Mao Zedong read the official histories of all 24 dynasties multiple times, which deeply shaped his political, military, and strategic thought. You find a lot of historical context and references in his formulation of strategic, military, and political thought—and his actions in leading the revolution toward success in the earlier days. In particular, he read multiple times the chapters about six battles in various dynasties where the weaker party initially retreated but eventually won through smart strategies. He internalized the learnings and navigated through the decades where his revolutionary forces were considerably weaker than the opponents.

History offers rich case studies on strategy and decision making: How did Liu Bang, a low-level bureaucrat turned warlord in the second century BC, go on to beat Xiang Yu, a much stronger rival who had money, power, connections, and family legitimacy, to found the Han Dynasty? Why did the illiterate first emperors of Ming Dynasty in 14–15 centuries do so well, but their much better educated, much more informed offspring fail miserably and eventually cede the country to the Manchus?

A *key lesson drawn from history is a deep understanding of people and human nature*—including how to rally people in uncertain situations, how to build alliances, and, once the war was won, how to distribute spoils and at the same time keep the subjects loyal and yet still motivated to build the new regime. It is the same for any founders building a startup today.

LIU BANG VERSUS XIANG YU: A STORY OF DAVID AND GOLIATH, AND MORE

At the end of the third century BC, as the Qin Dynasty was in terminal decline, China entered a period of chaos, where various warlords fought to reunify the country again.

Among the most promising contenders was nobleman Xiang Yu, descendant of a famous general who controlled much of China at the age of 25. Much older Liu Bang, who was born a commoner and held a post of a low-level bureaucrat before the chaos descended, was not seen as a hopeful at all.

However, Liu, less educated and less noble than Xiang, exhibited much greater political savvy and leadership. He had the correct understanding of the competitive landscape, put the right people on his team, and built allies and partnerships. In comparison, Xiang was less strategic, proud, often acted on impulse, and was often offensive toward allies and key advisors.

For example, when Liu took the Qin Capital, Xianyang, he was tempted to move into the palace and declare himself the new emperor. His advisor stopped him, counseling that it was too early to put himself on the opposite side of all the anti-Qin forces. Liu listened and instead appeased the citizens who had been oppressed by the Qin regime. When Xiang sacked the same city later on, he simply pillaged and massacred, out of impulse.

Within four years, Xiang lost the war and committed suicide. Liu went on to found the Han Dynasty, whose name is still used to describe the ethnic Chinese nationality. ■

IMPRINTS OF FOUNDERS OF CHINESE TECH FIRMS

In a disruptive environment, a good or a bad leader makes a big difference. Even a more aggressive versus a less aggressive leader, in different circumstances, can lead to very different outcomes. This is because things move very fast, and disruptors can be disrupted overnight. A strategic misstep or a wrong judgment could become fatal very quickly. "In traditional industries in China, even if you are in a terminal decline it will take months if not years for things to unravel. In tech, sudden death is quite common," a tech investor in China shared.

The situation startup leaders are facing is in a way similar to *Journey to the West*, where any of the 81 hardships could completely destroy the mission of Sanzhang and his team.

Throughout our career, over which we interacted with many Chinese tech leaders of various company sizes and levels of success, we realized that the successful ones, while on the surface can be very different, have the following traits in common:

- They are more often than not flamboyant but always inspiring.
- They are on the ground and *have suffered difficult circumstances* together with the team.
- *They are deeply reflective and sometimes philosophical.*
- They draw inspiration from historical events and personalities.
- They discuss and debate a lot with experts.
- They have a *profound understanding of human nature* and know how to distribute spoils of success and reward team members.
- They are passionate and determined—just like Tang Sanzang was in *Journey to the West.*

Let's take the examples of Lei Jun, founder of Xiaomi, and Duran Yongping, founder of BBK, the consumer electronics group that incubated smartphone brands Oppo, Vivo, and Realme. All the BBK brands and Xiaomi are successful in global markets outside China.

Lei, with his software background (he previously led both gaming and productivity suite companies), was an avid builder and operator of fan communities. Xiaomi's online-focused marketing tactics created euphoria among fans and defined its initial success. Lei, who often spoke at conferences, was even dubbed "Leibusi," or the Chinese Steve Jobs, for his style.

Lei often shares his reflections with the team as well as at public events. One of our favorites is: "Never use tactical diligence to cover for strategic laziness." He used this to point out a common problem facing leaders: getting so busy in minute operational details that they stopped learning, thereby depriving the organization the ability to grow even further.

In contrast, Duan was very quiet in public. He achieved his initial success copying Nintendo's gaming console in China. After founding BBK, he cleverly marketed gaming consoles as cheap computers for children to catch up with the upcoming computer age—which convinced many Chinese parents to open their wallets.

His major achievement then was actually building a nationwide channel distribution network, and the ability to manage and incentivize the channels as well as building a large on-the-ground sales force. That, coupled with the blanket offline marketing, was the playbook that defined the success of Oppo, Vivo, and many other brands.

Jet Lee, an Oppo executive sent to Indonesia, later founded J&T Express, which is now fast becoming one of the largest ecommerce logistics company in the world. Many other consumer and ecommerce brands were also incubated through the network, leveraging the initial framework Duan had created. Duan also mentored Pinduoduo's founder, Colin Huang, whom we discuss extensively later in this chapter.

Among our favorite quotes of Duan's is one on daring to be a latecomer, as we highlighted in Chapter 2: "Let you start first, and I will go for the kill with the model you have explored in a market you have educated. In your eyes, I am your competitor; in my eyes, you are my tool."

That quote was demonstrated in Oppo and Vivo's confidence in challenging Samsung in the global smartphone market and in Pinduoduo's daring attack on Alibaba's dominance in ecommerce in China.

Both leaders have left a lasting imprint not only on their organizations, but also on a whole suite of companies that they have incubated, inspired, and invested in.

Both have built successful companies where not only themselves but also a whole entourage of lieutenants became rich; both have, on numerous occasions, shared not only their success story, but the struggle they faced, and, more important, the reflections and learnings about people, organization, and products that they feel could benefit others.

Their companies bear deep imprints of Lei's and Duan's respective personal characteristics, background, and life experiences, as well as their cognitive framework. We see similar imprinting effects on other Chinese ecosystems of companies.

And in a way, you can also draw a parallel between these leaders and Steve Jobs, whose interest in calligraphy both at university and as a hobby deeply shaped how Apple designed their products and looked at aesthetics.

Decades of research have consistently shown that organizations are a reflection of their top executives' and CEOs' experiences, cognitions, and psychological orientation.[1]

The imprinting effect of the founders and CEOs on the organization can last long and gradually become the organizational DNA, because a series of strategic decisions reflecting the founder's characteristics can reinforce such an effect over time.

Such an imprinting effect also determines how effective or successful a company is in making a decision in response to external stimuli and shifts, such as the consumer touchpoint migration from PCs to smartphones in the early 2010s, the disruption triggered by the COVID-19 pandemic, and the entry of strong competitors into a company's market.

In fact, to understand why Chinese tech companies undertake certain strategies or to predict these companies' next actions, it is often more effective to analyze the characteristics of the leaders rather than the points of view of the market or financials.

 ## HOW JACK MA AND PONY MA OVERTOOK BAIDU'S ROBIN LI

Precisely because of the fast-changing nature of tech in China, leaders have to be constantly on their toes to navigate their organizations through (often challenging) external changes.

Internationally, there have been numerous case studies on why once successful organizations fail: Nokia, Motorola, BlackBerry, and Yahoo. If you dig a bit deeper, it is almost always the case that leaders fail to respond to external changes timely and decisively.

The same has happened in China. In the 2000s, as internet access expanded, people started doing more than just "surfing": Baidu, Alibaba, and Tencent (with the famous collective acronym BAT), emerged to serve the needs of search, ecommerce, and entertainment, respectively.

Their founders, Robin Li, Jack Ma, and Pony Ma, started with different backgrounds. Robin Li, with a degree in computer science from the University of Buffalo, was the star among the three and the bigger draw of tech talents. Pony Ma graduated from the barely known (even to this day) University of Shenzhen, also with a degree in computer science, and started his career as an engineer with a relatively unknown telco service provider. Jack Ma's personal stories were more well-known—he was trained as a teacher, and his first startup was a translation agency.

Fast-forward to 2021, Tencent and Alibaba are the towering giants in China's tech industry, with market caps of US$722 billion and US$615 billion, respectively, at the end of June 2021 (we used this milestone because it was right before the government crackdown sent tech stock prices into a nosedive). Baidu, however, fell behind big time.

In other words, Baidu, founded by a prestigious US returnee, fell behind the other two members of BAT, whose founders come from much humbler backgrounds.

Commentators often attribute this to the pride, lack of trust, and centralized decision making of Robin Li. However, after speaking with a number of veterans personally close to these people, we see a different picture.

Compared to Li, the Ma duo both experienced a lot of hardship and setbacks early in their careers and entrepreneurial journeys.

Pony Ma had to pretend to be a female user to chat with the early adopters of his instant messenger who couldn't find people to chat to; he had two business cards—one said General Manager and the other said Product Manager—which made some people believe he was a fraud; and when he could no longer afford the server costs of running the instant messenger, nobody was interested when he tried to solicit a buyer.

When Jack Ma first started China Yellow Pages, the precursor of Alibaba, in 1995, he had to ride his bicycle across Hangzhou to sell his online packages to businesses, many of which, having not even heard of the internet, thought he was a swindler. Even families of his earliest employees thought the same.

Both Mas had often-painful experiences working in the complex, and often brutal, Chinese business society, where trust is low and competition is fierce.

After buyers rejected Pony Ma's proposal to sell OICQ, Tencent's first instant messenger, he had no choice but to build services on top to try to monetize its product; when banks refused to provide escrow services for Alibaba's "small ticket" ecommerce transactions, Jack Ma had to build his own escrow, which eventually evolved to become Alipay.

There are many such stories in the Mas' entrepreneurial journeys.

As a result, while Jack is flamboyant and Pony soft-spoken, they share the same acute sense of the environment they are in and the people around them. That allowed Tencent and Alibaba to keep reinvigorating to stay ahead of disruption, thus far.

A key example is the fast proliferation of mobile internet access in the early 2010s—when smartphone penetration expanded rapidly in China. Alibaba poured resources into the mobile app version of Taobao as well as the Alipay app.

Tencent responded swiftly as well. It launched an internal competition to create a mobile social app. Three teams, isolated from each other, were racing to develop WeChat at the same time. Eventually, Allen Zhang's Guangzhou-based team won, beating its Chengdu-based internal competitor by one month.

More important, in order to support WeChat, Pony Ma decided to canni-balize QQ, its long-term instant messaging champion in the desktop era. He later reflected: "In the era of mobile internet, a company might seem invincible but actually has a lot of risks. If you do not grasp the trends in this society, you are in a very dangerous situation—whatever you have accumulated could crumble overnight."

Pony Ma knew that to constantly grasp new trends is not easy at all. There are internal vested interests (such as QQ's success), product DNA, and other factors in the organization. He says, "My way is to have another backup, open up an independent unit, that is unrelated to the current products or ways of doing things."

He also convinced the QQ team that it was necessary. "We have the shares of the same company, and therefore are on the same boat. So everyone needs to have a longer vision, instead of narrowly focusing on the interests of your department. I think people can understand that."

Interestingly, despite the fast-growing viral app WeChat, QQ later eventu-ally launched a mobile version as well, and shifted its focus to younger audi-ences. "Many young people do not want to be in the same social circle as their parents. They will be very active on QQ but use a completely different style to share anything on WeChat (which is used by their parents)."

Outside China, Google also realized the importance of mobile, especially how it will change the way people search for information. It went on to buy Android, and pay Apple tens of billions of dollars to keep its search engine and map as the default on iPhones. The dominance of Chrome as a mobile browser and Gmail as an email tool kept the users of the mobile age, as well as advertisers.

Baidu was supposed to be the Google of China, but it completely missed the mobile internet, while Alibaba and Tencent built their own walled gardens—complete ecosystems of subsidiaries and portfolio companies (for more infor-mation on tech walled gardens in China, see Chapter 7, Product).

Ex-executives we spoke to pointed to Robin Li's hesitancy and lack of urgency—desktop search ads were the biggest revenue drivers for Baidu, with the corresponding business unit holding a huge sway over the company's deci-sion making.

Like the Monkey King, Robin Li was very confident about the company's existing technology and products (and indeed he has reasons to be confident in the core technology). He later recounted that at the time he did not think mobile search was much different from desktop search, aside from the slow speed (as 3G technology was just being rolled out).

Much later, in a letter to investors in 2021, Robin Li admitted that the previous 10 years had been the fastest-growing era of mobile internet, with countless business model innovations. He said, "We discovered that the technology we liked and were good at seemed to not function anymore, and that caught us off guard."

Baidu tried to invest in a number of new business areas, including food delivery and artificial intelligence, to diversify from search. Li brought in successive high-profile executives, including AI guru Andrew Ng and Microsoft veteran Qi Lu, none of whom lasted long. To this day, there has been no change and search ads are still the dominant revenue sources for Baidu.

An ex-senior executive shared that a key reason Baidu could not succeed in new business areas was because Li's mindset was fundamentally alien to China's complex business reality, and his understanding of people was simplistic. "Otherwise, how would he not be able to rein in the search ads unit?"

It seems to be no different from Amazon building cloud (because it can) and video streaming (because it needs to retain ecommerce customers). However, Alibaba's approach is more than using its scale to flip a cost center to revenue center; it's more a matter of responding to a problem that prevents its growth, while keeping current and potential competitors at bay.

How did business leaders sense these new opportunities? By firmly being on the ground and relentlessly talking to various stakeholders—resting on the laurels of successful experiences and being detached from reality were the biggest risks that Chairman Mao had highlighted earlier.

THE (INTERNATIONAL-MINDED) NEW GENERATION

Later founders, including Wang Xing of Meituan, Zhang Yiming of ByteDance, and Colin Huang of Pinduoduo, did something similar—being on the ground to understand the fast-changing Chinese market and lead their firms to respond promptly to the needs of their clients. Once they realize there is a better opportunity within their grasp, they move very quickly.

Fundamentally, Wang, Zhang, and Huang are *thoughtful strategists, relentless executioners, and resilient survivors*. They are thoughtful with their idealism, setting the vision for the firm, especially when the firms were still in the dark tunnel fighting for the future without a clear direction. They are resilient survivors, being serial entrepreneurs and market creators for multiple times, trying every possibility to survive, continuously facing the threat of death. They are also pragmatic and relentless executioners. They are determined to push forward and advance despite all the hurdles.

What's more, they are young, energetic, and engineering-trained. They are all deeply influenced by the entrepreneurial culture of Silicon Valley, making them produce the potential of international companies.

Huang studied in the United States and worked for Google; Wang had a brief stint in the United States after graduating from Tsinghua University, but dropped out to start companies in China; and although Zhang did not study or work in the United States, he talked a number of times about his reading of Silicon Valley companies and their culture(s). Among these are Netflix's "context, not control" philosophy and Google's OKRs (objectives and key results)—the latter of which is probably more rigorously adopted at ByteDance than even at Google.

Like their forebears, the trio had a string of entrepreneurial experiences, including small successes and notable failures, before starting their current companies and making them big. Coincidentally or not, all three had exactly seven years of entrepreneurial experiences before founding their respective current companies.

Meituan

Wang Xing, the quiet, thoughtful founder of Meituan, had many failures under his belt before starting Meituan. He quit his postgraduate studies in the United States to go back to China to start companies in 2003, with his initial string of startups copying ideas from Silicon Valley (including a Facebook-like social network and a Twitter-like microblogging service). Many ridiculed him as a simple copycat, until his Groupon copy Meituan, founded in 2010 (seven years after his first startup), started to gain traction.

However, even during the eight years from Meituan's founding to its IPO in 2018, investors were divided about Wang's business. Some doubted his business model: "The huge costs of building and maintaining the expensive delivery network will never pay off." Others are more sanguine: "If someone can make it work, it's gotta be Wang Xing."

Every sector that Meituan participated in was fiercely competitive—from group buying at Meituan's founding, where more than 5,000 companies join the fight, to food delivery, where Meituan fought hard against well-capitalized Baidu and Alibaba-backed Ele.me, to online travel, where Ctrip had built a seemingly impenetrable fortress, to the recent community group buying, a messy war against almost every consumer internet giant in China: Pinduoduo, Alibaba, JD.com, Didi—Meituan's wars seem to be never over.

From battle to battle, Wang built deep trust among his top lieutenants. He once said: "Things in nature do not have clear boundaries, therefore we should not set boundaries to ourselves. As long as the core is clear—who do we serve? What services do we provide? We will keep attempting new business areas."

Some people have described Wang as a cockroach, that is, he never stops fighting. Others say that he crawled out of fields of the dead. People who are close to him describe him as a man of few words but very deep thinking.

A few key decisions he made after careful deliberation helped Meituan avoid cost traps: an early example was when many large group buying companies were pivoting to not only services but also goods to keep growing, Wang decided to not do so. It turned out that these companies were quickly obliterated by Alibaba after launching its own group buying services for goods. Later on, even as Meituan took an investment from Alibaba to survive, Wang engineered a bigger investment from Alibaba's arch rival Tencent to stay independent, unlike its competitor ele.me, which was eventually absorbed by Alibaba with its founder pushed out.

ByteDance

Zhang Yiming of ByteDance and Colin Huang of Pinduoduo, another two of the new age empire builders, are both known to be deep thinking and ruthless in execution and for driving their people really hard.

Coincidentally, Zhang came from the same city (Longyan) in Fujian province that Wang did. After graduating with a degree in software engineering in 2005, Zhang embarked on a journey of entrepreneurship. By the time he started Toutiao (the early name of ByteDance), a news aggregator, in 2012, Zhang had been an entrepreneur for seven years. It is worth mentioning that most of his previous entrepreneurial projects were related to "recommendation algorithms," which provided a solid foundation for Toutiao's news (and later TikTok's video) recommendation engine.

A series of early work experiences on recommendation algorithms have deeply influenced Zhang's later strategic choice, and has been the direction that ByteDance has always adhered to.

ZHANG YIMING AND HIS STUDY TRIPS

Interestingly, Zhang Yiming didn't have any actual study or working experience outside China. Instead, he spent a lot of time traveling and learning about the overseas market.

Dreaming of ByteDance being a globally competitive company, Zhang realized that the first step was for him to learn English well. It was reported that he not only spent a lot of time practicing English, he also asked the company's top management to learn.

In September 2014, Zhang first paid a visit to Silicon Valley. He visited Facebook, Google, Airbnb, and tried out Tesla's new model. The products, organization, and management of these international companies left a deep imprint on him.

After the Silicon Valley trip, Zhang accelerated his "study trips" abroad. He made two consecutive trips to Japan in 2015. In the same year, he was invited to the United States to participate in the Sino-US Internet Forum. In 2016, Zhang and his early backer, Cao Yi, set off on an investigation trip to Delhi, Mumbai, and Bangalore, India.

Studying abroad has undoubtedly increased Zhang's understanding of overseas markets. ■

Pinduoduo

With a master's degree in computer science from the University of Wisconsin, Colin Huang was lucky to meet William Ding, founder of NetEase (and one of the successful first-generation tech entrepreneurs in China) early on, who referred him to Duan Yongping of BBK, mentioned earlier in this chapter. Duan, deeply impressed by Huang, brought him to meet Warren Buffett in 2006.

Seeing the "glass roof" in working in multinational corporations, Huang left Google to become an entrepreneur in 2007, building a series of companies in sectors including mobile commerce, ecommerce enabling, and gaming. While people were surprised at Pinduoduo's rapid ascent after its founding in 2015, few noticed that Huang already had been building companies for more than seven years (the same as Zhang and Wang), with a string of minor successes and failures that built his understanding of how Pinduoduo should take shape.

As we can see, founders of these new generations of giant companies exhibit a strong combination of international inspiration and social/entrepreneurial experience. Their success over competitors is not purely because of luck, but from their training from years (and companies) of startup experiences.

Without a doubt, Meituan, Pinduoduo, and ByteDance have shaken up the dominance of BAT in China. Not all of them are overseas yet, but building global companies is deeply rooted in this generation of founders and, like ByteDance, once a decision has been made, the execution in expansion is relentless.

 ## LEADERSHIP IN GLOBAL EXPANSION

A leader of a consumer tech giant in China once shared with us that building a business overseas is like "the second entrepreneurial journey," and it was certainly more difficult than the first one in China. The biggest factor, he said, was that "We still have our business in China—and it's damn competitive here."

Another tech leader shared with us that "suddenly I am blind!" when he was expanding overseas.

These two points of view are common among many Chinese leaders expanding overseas. Why?

In China, you are in the market, you are leading this. And you have to figure things out. And you have a good sense of the market. You know who your competitors are, you know who your customers are, you know what products you are building, and you have one single market to deal with. But when you go outside China, suddenly there are all these different dimensions you need to figure out. And you are doing this while the market in China, which is throat-cutting competitive, is demanding a lot of your mental space and attention.

International markets are more complicated or new, and there are more dimensions to deciding which market you go to. How much do you need to understand about each market? While this is something that all companies going overseas will face, there is a different pace for tech companies. Traditional companies have the luxury of time—maybe 10 years to figure out a market. But for tech companies a lot of things need to be figured out in a much shorter time. Make or break happens much faster. And Chinese companies have the baggage of a successful (and highly competitive) home market that they need to pay attention to.

How does this differ from American companies that also have a large and successful home market? Many American companies have two natural advantages when they expand overseas. First, early adopters in their overseas markets are typically Western educated. For instance, the first group of users of Facebook in Southeast Asia or Middle East were those who graduated from US schools, and they were also the local trendsetters. Second, American firms have a talent base that they can tap into from these markets who are, again, Western educated. These talents can align themselves to the culture of Western tech companies. Chinese companies expanding overseas have neither.

In China, when tech companies try to figure out a strategy, they usually do not use consultants. Instead, they personally spend time talking to experts, market participants, and other stakeholders—mainly through introductions and networking in friend circles (*guan xi*). The experts, who are also Chinese, can easily relate to the key questions and concerns of these leaders.

This is different outside of China.

"When I go to Indonesia, where can I find the right people to talk to? And when we speak in our nonnative languages, how much real message is lost in a conversation?" the tech leader said. "I just can't build the same level of quick grasp of a market as I have done in China."

With competing businesses in China that also require their mental attention, leaders are just not able to spend as much time and effort on the ground in a foreign market as they did when they built up the company in China.

This is why it is much harder for Chinese companies to succeed outside China, and for those that have done so, they are a force to be reckoned with.

We think there are three critical elements that these Chinese leaders and companies should have in order to succeed overseas.

Mental Space and Judgment: Juggling Demanding Chinese Markets with the Complexity of New Markets

In a new market with incomplete information, it is often better to use existing recipes as a starting point and take an iterative approach when the market gives you the feedback to make adjustments—this is tech startup 101. This is how successful Chinese tech leaders expand their businesses into new segments in the China domestic market. Outside of China it has proven to be much harder because the leader is not there all the time, and therefore the right feedback might not be able to resonate in a timely way that it would for local competitors. Western tech companies had this same challenge when they entered China, which led many to fail.

Daniel Zhang (no relation to the ByteDance founder), Jack Ma's successor as chairman of Alibaba, built his career success in Alibaba when he served as the president of Tmall, and created the Singles' Day (November 11) shopping holiday. Nowadays the annual shopping event has generated gross sales three times higher than Black Friday and Cyber Monday combined! The Singles' Day under Tmall has made Zhang's reputation and paved his way to get the CEO baton from Jack Ma.

While Singles' Day has now become the biggest shopping festival in Southeast Asia, in Zhang's mind, Tmall's business model (with a main focus on brands) was also how a successful ecommerce business should run from the start.

This inevitably impacted his thinking about the strategy for Lazada, Alibaba's ecommerce subsidiary in Southeast Asia. One executive we spoke to said that Zhang focused too much attention on making Lazada Tmall-like—perhaps a reflection of his Tmall background in China.

In the meantime, Lazada's regional competitor Shopee, which perhaps understood the overall market structure and, crucially, its stage of development, in Southeast Asia better, decided to focus on a C2C marketplace to start. Interestingly, this is similar to how Alibaba started their own C2C marketplace, Taobao, back to 2003.

Even if the Tmall model would eventually prove to be successful in Southeast Asia, Lazada would have lost the momentum of accumulating customers and expanding its support infrastructure in Southeast Asia. However, if Shopee decides to focus on its B2C outfit ShopeeMall, it would be able to do it much more easily, just like Taobao launched Tmall in China in 2008.

When leaders make strategic decisions, it is natural for them to be influenced by their experiences, especially successful ones. Developing new mental frameworks to incorporate the new inputs in a different market is one of the key challenges for any leader. For Chinese leaders who have developed very customized mental frameworks based on the specifics of the Chinese market, it can be even more challenging.

Running both Chinese and overseas businesses at the same time could further complicate this, because the demanding market reality in China subconsciously would reinforce the preexisting mindset, even though there is new feedback from global markets.

Therefore, it was no surprise that Yiming Zhang of ByteDance wanted to stop running the day-to-day operations in China in early 2020 to focus on how to make TikTok a truly global company. He would need that mental space to make it work.

PATH DEPENDENCE

Organizations could be an imprint of their leaders, whose cognitive framework is based on their own characteristics and background. The way they understand the market to develop the strategy and execution is influenced by their prior experiences.

When leaders' cognition is based on success in the domestic Chinese market, they could become more confident in the existing (i.e., "proven") recipe, and are less flexible or open to different views in overseas markets. ■

Commitment: How Many Resources Should You Pour into New Markets?

We often see a dilemma for many Chinese tech companies: while founders/ CEOs are often keen to develop international markets, China's domestic market is always the core. Besides, global markets are fragmented, with each single market occupying perhaps an even tinier percentage of the potential. They might nonetheless require a similar level of attention compared to larger markets including the home market of China.

The commitment comes in different forms: funding, people, organizational support, and leadership attention. More important, when encountering challenges in the new markets, would the leader double down on the commitment to resolve the challenge, or not?

THE DOMESTIC MARKET IS STILL THE CORE

Didi's IPO prospectus revealed that after years of development, international business accounted for only 2% of the company's overall turnover. By contrast, 43% of Uber's revenue is from regions outside the United States and Canada.

Similarly, Alibaba's international commerce (including retail and wholesale) in FY2020 was only about 7% of the company's total revenue. That included its cross-border business, AliExpress, which did not have substantial actual operations outside China. ■

To develop international markets seriously, leaders need to justify all the investment and resource allocation by market potential. In addition, leaders also need to steer the organization toward supporting such initiatives when domestic business units will naturally hold much more weight. Finally, leaders will need to make sure the people working on international initiatives feel motivated and supported—we have seen many expansions fail not because of market reality, but because the best managers will not want to be on such projects versus domestic ones which are perceived to be better for their career (this is discussed more in Chapter 5, People).

It is a chicken-and-egg problem—if you do not pay an outsized proportion of attention, it is quite difficult to raise the low percentage of the international markets' contribution to total revenue as well.

The competition between Kuaishou and ByteDance internationally in the small video app space clearly shows the challenges.

Kuaishou

Kuaishou started its overseas business at about the same time as ByteDance's TikTok, or even a little earlier. However, today, while TikTok is one of the most popular apps in the world (and has the attention of US politicians), Kuaishou's international business is present in just a few pockets of markets, in almost all of which TikTok is bigger.

Ex-executives who were part of Kuaishou's international efforts shared that the divergence in the companies' performance lies in the resources invested from headquarters, which was influenced by the leaders' judgment, understanding of the overseas markets, and eventually their commitment.

Kuaishou's first serious international attempt started with testing a few key markets with localized apps and customer acquisition campaigns. Markets that responded well, including those in Turkey and Brazil, were then chosen as focus markets, where more resources were invested. Kuaishou's CEO, Su Hua, hoped that these international markets would grow on their own.

This is a standard approach used by Chinese gaming companies to launch their products overseas. However, video apps are different because they require local operations, especially in engaging local content producers, dealing with regulators, and developing advertisers.

Kuaishou also missed a critical step, which was analyzing why certain markets responded well to initial tests. As a result, when they encountered challenges in these markets, they did not have enough information to decide what to do next.

Facing such circumstances, Su Hua called a halt to the investment. He chose to not personally spend time on the ground to fix the issues and push for growth, but to focus on the China market. Nobody would want to be trapped in this money-losing business, especially when they do not know why it is losing money in the first place.

What Kuaishou experienced was very common: leadership decides on a direction, the response is not as good as expected, and leadership then decides to make a U-turn without spending enough time on the market to figure out the reason for the poor response.

The same thing happened to JD.com, Tencent, Ant Group (multiple times), and even smaller companies such as JollyChic.

ByteDance

In comparison, ByteDance is led by a founder who never hides his global ambitions. His aggressive acquisition of muslic.ly (a Shanghai-based lip-syncing app popular with Western youth) against other bidders, including Facebook, has demonstrated his deep international commitment. In an interview in 2018, Zhang Yiming said "our vision is to be a 'global creation and communication platform,' and we hope that is a universal platform."

Chinese industry watchers in Brazil shared that Kuaishou's initial success in Brazil led Zhang to almost immediately ramp up investment in that country. The rationales were that, first, it shows that short-form video has promise in Brazil; and second, ByteDance can't afford for Kuaishou to gain an upper hand in international markets.

Only this time, compared to Kuaishou, ByteDance was less opportunistic but more serious and committed.

In entering the Brazilian market, Zhang's series of actions clearly showed his determination. Brazil is the largest country in South America and home to more than 200 million people. On average, Brazilian consumers spend 9 hours 29 minutes per day on the internet. For Zhang, this is a battle he cannot lose if he wants to claim ByteDance (TikTok) as a global company.

In 2020, Brazil became the fastest-growing market for TikTok, and is the second-largest international market in user numbers. Zhang's constant commitment paid off elsewhere too, with TikTok ranked top in app stores in many countries.

Even facing the ban in India and pressure to sell from the Trump administration in the United States, Zhang soldiered on. In 2021, he showed the same commitment to ecommerce in Indonesia, even though the market performed very poorly for most of the year.

Opportunism versus Commitment

The preceding two cases and comparison show that many leaders still look at international markets in an opportunistic way. Despite the market uncertainty, TikTok poured significant resources into Brazil in 2018 under Zhang's leadership, renting and expanding office space in Vila Olimpia, São Paulo, the premium location for headquarters to many international tech companies. It further committed to hiring more people to expand the operation, host high-profile events to attract new users, and develop communities to create high-quality original content to share on the app. Unlike Kuaishou, Zhang didn't show any sign of hesitancy.

The challenge of overseas expansion, or the second entrepreneurship journey, is that international expansion is often dispensable and can be put on the back burner if other priorities take precedence. In contrast, in their first entrepreneurship journey back to earlier days, these two leaders had no choice but to carry on until success.

Lack of investment and commitment can result in a vicious circle, not only a drop in or halt of momentum, but also a departure of good talent who cannot see the future inside the company. For instance, Liu Xinhua, the former head of international operations at ByteDance, poached by Kuaishou and one of the very few senior executives at Kuaishou who spoke English fluently, left the company. The downsizing of human capital specialized in overseas markets leads to further decline in performance. Kuaishou intended to ramp up its international efforts again later, but the timing would already be very different, largely because of TikTok.

Leadership's Managerial Attention: Too Little or Too Much?

The importance of commitment is easy to understand; the question of leadership's managerial attention is a bit trickier. Shall the leaders pay close attention to the overseas expansion? The first reaction is yes, they should. Daniel Zhang visited Singapore once every month to look at Lazada before the pandemic, showing that Lazada and Southeast Asia is his and Alibaba's top priority. When Uber competed with Didi, its founder, Travis Kalanick, traveled to China much more often.

However, we should not equate leaders' attention to continuous hands-on management.

In 2021, Xiaomi ranked first in the Indian smartphone market with a more than 25% share, ahead of Samsung, Vivo, Realme, and Oppo. What drives Xiaomi's success in India? That it's cheap and easy to use, or its leaders' commitment?

Yes, both count. In addition, Xiaomi founder Lei Jun's efforts and attention in the Indian market must be considered. Lei visited India as early as 2001 and kept a diary of his visits. Over the years he had a comprehensive understanding of India's economy, culture, and software industry dynamics. In 2015, Lei attended Xiaomi's overseas launch event in India. His awkward English greeting "Are you okay?" went viral and made Xiaomi even more well known. When Lei visited India in 2017, Lei was granted an interview by Indian Prime Minister Narendra Modi.

What's more, while Lei paid sufficient attention to the India market, as he once advocated that the India market will be allocated with a higher priority for development than China's domestic market, he is willing to give Indian leadership a strong supporting hand. Instead of making all decisions by himself, Lei empowered his local leadership team in India on a number of critical steps.

As a result, Xiaomi product design in India meets the needs of Indian consumers according to the preference of the local market. To adapt to India's hot climate, Xiaomi adds more cooling modules to its phones; considering the wetter climate in India, Xiaomi put a thicker coating on the ports and charging cables; and so on. Localization of the product's features is a reflection of leaders' delegating more decision making to local executives, and trying to avoid too-detailed hands-on intervention.

In addition, Xiaomi has built seven factories in India! Considering the strong manufacturing capacities and easy access to the full supply chain in China, to relocate part of the smartphone manufacturer to India needs not only courage and commitment, but also the empowering of the local leadership teams.

XIAOMI'S SEVEN INDIAN FACTORIES

In 2015, Xiaomi's first factory was established in partnership with Foxconn in India. In March 2017, Xiaomi and Foxconn cooperated again to build a second factory in Andhra Pradesh. In November 2017, Xiaomi announced a partnership with India's Hipad Technology to produce charging banks and set up its first mobile power plant in Noida.

In April 2018, Xiaomi announced three new factories at the same time. They localize the management and have only one Chinese member on the executive team.

In 2019, Xiaomi set up a new handset manufacturing plant in Tamil Nadu, covering an area of about 93,000 square meters. These localization and leadership measures help Xiaomi sustain its position and grow during the China–India conflicts when many Chinese firms are banned in India. ■

Although the long-term benefits of the Indian market for Xiaomi remain to be seen, the fact that it managed to break through a local market and talent ecosystem shows its commitment. Lei tried to strike a balance between leadership attention at the headquarters and empowerment or hands-off management at the local level.

However, paying leadership attention to the overseas market but resisting the temptation of daily involvement in the operational details is not easy for leaders, especially for the first generation of founders of Chinese tech firms.

If the CEO pays too little attention to the foreign market, not only will the company miss many opportunities, but it can also lead to poor operating performance, especially in a competitive environment.

Often, on the one hand, committed leaders who spend too much attention will end up interfering with the functioning of local leaders and sow distrust. On the other hand, too little managerial attention from leadership builds up misunderstandings.

The key is not how much time the leader spends on the ground, but rather, whether they can build up trust and rally the right people on the ground to run in the same direction.

Throughout Chinese history, there was a deep mistrust of processes and institutions, especially at the higher level. Historical leaders were successful not because they built the best systems, but because they pulled the right strings of people around them: reward them for their achievements, punish them for major mischief, tolerate and even make use of small mistakes or flaws, and, most importantly, balance their desires with those of their peers.

One Chinese tech leader once shared, "For the key people around you to work well with you, you need to deeply understand their motivations, and distribute both hardships and rewards accordingly." He believed that ruling by processes, while seemingly fair, can lead to misalignment of objectives among the top echelons, especially in the fast-changing environment.

Modern experts claim that leading through a human-centered approach, rather than processes and institutions, was outdated and a chief contributor to China's decline in the eighteenth and nineteenth centuries. However, this mindset still prevails in China, from the top political leadership to small and mid-size enterprises (SMEs). Something that is deeply rooted in history will die hard.

 ## QUESTIONS TO LEADERS

We have witnessed far more turbulence than smooth sailing in Chinese tech firms' global expansion. The undertaking is inherently difficult, just as most tech leaders' entrepreneurial journey was in the first place (remember that founders of ByteDance, Meituan, and Pinduoduo all spent seven years building and failing before founding their current company).

While we will explore people, organization, and product in the next chapters, leadership is ultimately responsible for all these aspects, and for the success or failure in overseas expansion.

Before undertaking any global expansion efforts, leaders need to have a clear understanding of themselves and their organizations, as well as the commitment they are willing to make.

For the localization requirements of the destination markets, it is important to understand that things (including timing) might be very different from those in China, spend as much time as feasible in the market to understand the norms and nuances, and identify the right people and organizational structure to lead such expansion efforts.

Leaders also need to have the courage to decide against any rushed international efforts if they do not have the right leadership attention, right people, right organizational capabilities, or the right timing in place.

Wang Xing of Meituan spent some time, at the beginning of 2020, in Indonesia's capital Jakarta, to figure out what the market is like and whether Meituan should consider a role in the market. Meituan had already made an investment in Gojek, the leading local ride hailing, food delivery, and digital payment in Indonesia.

During his month in Jakarta, Wang consulted various stakeholders and industry experts and floated multiple ideas, including taking a majority stake of Gojek. His final decision was to not yet enter Indonesia, but to focus on business in China first, as competition for community group buy was intensifying. It turned out to be a good move in light of the COVID-19 pandemic.

Leaders who decide not to pursue some seemingly promising international markets for the time being should remember what BBK's Duan said: dare to be a latecomer.

To conclude this chapter, we would like to raise a few questions for leaders who are thinking of venturing overseas.

- ▦ How open-minded are you? What could be the potential knowledge and mental blind spots?
- ▦ Do you have the mental space to deal with the challenges that come along in foreign markets, and make the decisions necessary?
- ▦ What is the purpose of your overseas expansion? How would you deem it successful? How committed are you to reach that success?
- ▦ How many resources are you willing to invest in such expansion—and what would be the decision process if the business performance turns out to be very different from the original expectation?

- How much time and attention can you possibly spend on the destination markets? How do you make sure you get continuous good information from the market to make the right judgments?
- Does your company have the right capabilities to take the expansion to success? If not, how do you, as a leader, get the capabilities ready? By self-development, alliance, or acquisition?

NOTE

1. D. Liu, G. Fisher, and G. Chen, "CEO Attributes and Firm Performance: A Sequential Mediation Process Model," *Academy of Management Annals* 12, no. 2 (2018): 789–816.

CHAPTER FIVE

People

Trust makes everything simple.

—*One of six core values of Alibaba Group*

A LTHOUGH LEADERSHIP IS CRITICAL for an organization, the leader alone will not be able to build a successful tech business. They need good people and the right people around, to not only execute but also to brainstorm solutions, and advise the leader, as well as sometimes even boost the morale of the CEO.

Undoubtedly, people are the core of the organization.

Especially in international markets, when leaders do not have the mental space and capacity to make all the decisions and deal with all the nuances, strong lieutenants are needed to help them operate the market, tackle challenges, and eventually build successful expansion.

This is the same situation the leaders had in China during their first entrepreneurship journey: Alibaba would not have been successful in the first place if Jack Ma had not had his 17 other cofounders; Wang Xing would not have built Meituan to where it is today without his top lieutenants, including Wang Huiwen (no relation) and Mu Rongjun sticking with him.

Any tech startup in China that eventually became a resounding success started like a guerrilla force fighting a war in their early days. During that time, they could not pay people the best money or provide the best working environments. They could not offer benefits, and most often they were not able to attract the best or smartest people in the market, for the right reason.

How would they find the people who fit the organization, people who were motivated with the vision, people who would roll their sleeves up and join the same journey? Why would Joe Tsai and Martin Lau give up their lucrative jobs in Hong Kong to join Alibaba and Tencent, respectively, in the companies' early days?

Every leader would agree that throughout an organization, people are the most important asset, the most important driver for the organization to build success after success, but can also be the most significant drag if not managed properly.

While Silicon Valley tech companies have a whole support network to tap into—experienced managers to hire, consultants and headhunters to use, market research to help with decision making, and SaaS tools to solve individual processes—most Chinese tech firms had to figure things out by themselves. Many were just like Mao's early guerrilla forces: "Ten plus people, with seven maybe eight rifles" or "rifles with bags of millet." There was nothing to fall back on except the will and determination of the people in the fight.

People issues are present throughout the lifespan of any organization. As Chinese tech companies grow and become more mature, things will evolve. When more experienced and better qualified managers are introduced, sometimes to take over portfolios from the more grassroots early team members, how do companies balance the interests of these two groups of people to move the organization forward coherently? How do they build affinity among the new hires, keeping the earlier revolutionaries from wearing out in the meantime?

Tech companies in China now are among the most attractive employers, even beyond banking and western multinational corporations (MNCs). Young people, who have formed the backbone of new blood for tech companies' growth throughout their history, now have very different incentives and motivations compared to prior generations one or two decades ago. Leaders who still preach hardship and endurance, a message that resonated 10–20 years ago, are now met with skepticism and resentment from the current young generation who grew up in an age of relative abundance.

Government has also recently stepped in to intervene in some people management practices, for example, discouraging overtime (996, 9 a.m. to 9 p.m., six days a week) so that young people might have more children.

When companies expand overseas, they encounter another set of people challenges: Should they send their loyal lieutenants who are more familiar with the organization, or hire local experts who are more familiar with the market? How do they marry these two different skillsets that are equally important for success? Do they fit the local senior hires into the company's culture, or adjust the culture to something of a hybrid? If the latter, how do they get existing people to adjust their approach to the new hybrid culture?

And, if loyal lieutenants are sent to foreign markets, how do companies make sure they become pioneers, not marginalized exiles? After all, in Chinese history there were plenty of promising mandarins sent to and then forgotten at the frontiers.

In this chapter, we cover all these elements through our conversations and learnings from Alibaba, ByteDance, Meituan, Pinduoduo, and others, from early days up to the present as they venture overseas.

FROM SCRAPPY GUERRILLA TEAMS TO TOP CHOICES FOR BEST GRADUATES: EVOLUTION OF PEOPLE IN CHINESE INTERNET FIRMS

When we talk about people, first and foremost is identifying the right people at each stage of the company's development, and bringing these people onboard.

We all know that large tech companies, including Chinese internet firms, nowadays have high hiring standards. However, we are less interested in the hiring process and standards and more in what type of talent suit these Chinese tech companies at different stages, and how to find such talent.

It is an evolving process—*finding the matching talent for each, and subsequent, phase of the company's growth.*

Zhang Yiming, who made the pledge to make ByteDance a global company at the beginning of 2020, has been known in China for attaching great importance to having great talent in his organization. In multiple speeches and memos, he described his recruitment philosophy as "do challenging things with great people," that is, identifying candidates with strategic thinking, shared values, well-rounded knowledge, and problem-solving abilities, to match with "challenging things."

Zhang's people policy is largely consistent with the fundamental tenets of the talent-management philosophy at Netflix put forward by Patty McCord in her book, *Powerful: Building a Culture of Freedom and Responsibility.*

Rifles with Bags of Millet: The Early Days

When Alibaba first started over 20 years ago, the hiring standard seemed to be surprisingly low. As one of the very early employees shared, "Anyone with decent looks, some matching skills, and the willingness to join the team, would probably get a job."

Among the original 18 cofounders of Alibaba, only Joe Tsai, a Yale graduate and former private equity investor, had a professional or credible background. The others all came from less-known schools and would otherwise have been destined to mediocre job and life prospects.

Similar things happened at Tencent. Among the notable early employees of Tencent, aside from Stanford-educated ex-Goldman Sachs banker Martin Lau, nobody had a glamorous background.

Chinese internet firms had to face the reality of the potential talent pool they could tap into, which was indeed very limited at the beginning.

Why? Because when Alibaba and Tencent were founded, China was in the very early days of the economic boom, and the top talent had plenty of choices. FMCG (fast-moving consumer goods) companies like P&G, as well as auditing firms such as PWC and KPMG, were the top choices of young ambitious graduates.

For computer science graduates, IBM, Oracle, and Microsoft offered better prospects than those scrappy entrepreneurial teams crammed into apartment units. Unfortunately, not many people owned cars or big houses in China back then, so the garage founder culture never took off in China.

However, it was exactly the scrappy teams that did wonders. They did not have the baggage of expectations from their university background or their family; they were hungry for success; and they were not distracted with the abundance of choices available.

What was left was an almost-single-minded focus on building the business, as a team. As we mentioned in Part One, people had no problem fixing issues at midnight or staying over the weekend. Many we spoke to said that rather than being driven by a mission to change the world, they were just hungry for success, and trusted the leaders who brought them on the right journey.

It is the relentless pursuit, driven by sheer willpower and hunger for success, that led these companies to gain their initial traction.

Cannons Replacing Shotguns: The Introduction of the Professionals

However, early generations of Chinese entrepreneurs, tech or nontech, realized very quickly that as their company grew to a certain stage, they would need more qualified and experienced talent to take their companies to the next level.

Joe Tsai built the funding and tax structure that allowed Alibaba to tap into more investment to further fuel the growth, while Savio Kwan, with three decades of MNC experience, joined as chief operating officer in 2001 to help Alibaba build the operations structure for scaling. Similarly, Martin Lau and James Mitchell, a former Goldman Sachs banker, proved indispensable to Tencent's growth.

However, putting external senior hires in charge is never an easy journey. Savio Kwan's early days in Alibaba were full of conflicts with the existing team, which felt that the suit-wearing executive and the management practices he brought in were too "fluffy."

Not every external senior hire survived—Qi Lu, former executive vice president of Microsoft, joined Baidu in 2017 as a high-profile case but lasted barely more than a year, before resigning "for health and family reasons." We discuss more about the dynamics between founding teams and external senior hires later in the section "Senior Ranks: Those Who Matter the Most."

The Elite Force: When Tech Companies Become Top Choices

While Chinese talents initially preferred MNCs, from the 2010s onward, the balance tilted toward tech and the internet firms. Young people, including graduates of top universities such as Tsinghua and Peking University, fight to get into the top tech firms.

There are several reasons for this:

- Tech companies are becoming ubiquitous and impacting every aspect of people's lives in China; as a result, young people no longer need to convince their parents and relatives of what they actually do (there could be a real stigma and social pressure if you graduate from a good university but end up in an obscure profession or company).
- Tech companies pay well—especially since Alibaba went public in the New York Stock Exchange in 2014, making thousands of employees millionaires, with a few becoming billionaires. People realized that companies

like Tencent, Meituan, and so forth are minting millionaires faster, and at a bigger scale, than their counterparts in other industries.

■ As the tech sector as a whole grows and competition intensifies, it needs many more talented people. Better pay and other perks, in addition to the equity upside just mentioned, attract the young and talented.

People also realized that in a fast-growing sector, their ability to learn, grow, and move up the ladder is accelerated.

While there is no shortage of applicants, top talent remains in short(er) supply. The competition for best talent, from fresh graduates to experienced hires, has also become intense, a reflection of the business competition among the companies as well.

ByteDance's Zhang has mentioned that candidates for roles requiring experience need to have qualities including *critical thinking, rationality, ambition, and self-control.*

EXTRINSIC MOTIVATION

Extrinsic motivation, which is often through monetary and other financial rewards, is often what the employees need. In big cities such as Beijing, Shanghai, Shenzhen, and Hangzhou where these internet companies are headquartered, young graduates face increasing pressure to buy a small apartment due to increasing real estate prices in the past decade. Owning a home in China is a social status symbol, a prerequisite for mothers-in-law to accept her daughter's fiancé and to agree on the marriage.

In 2008, a friend of mine who had joined Huawei a year before ended up in a large, less-developed country in central Africa. "Life is really tough here, people get sick and I can't find good food, at all."

"Why did you go there?" I asked.

"I get better compensation, as well as faster promotion," he said. "These are enough for me to be here."

He subsequently moved through a number of African countries. When I met a Huawei executive in 2019 at a dinner, I told him that my classmate was also in Huawei.

After checking through the internal directory, he exclaimed "Holy cow, he is six ranks higher than me, despite having the same years of experience."

So the decade spent in Africa was not in vain. My friend was indeed generously compensated and promoted. He bought a good house in Shenzhen, although he was rarely there, and his kids are receiving the best education in the city. ■

All the major tech companies dedicate significant money and resources to recruitment marketing and employer branding. Many are also hiring an army of recruitment agencies to help them fill positions.

A good friend of ours, who has worked as a data scientist at a top tech company for more than seven years, says that he receives on average 10 to 15 recruiter outreaches a week, even during the pandemic. Interestingly, his biggest challenge is also finding the right people for his Series-C funded AI startup, before his competitors and major tech firms. Part of what he does every day is outreach using Boss Zhipin, a recruitment tool that allows hiring managers to interact directly with candidates, bypassing recruiters or even internal HR.

ByteDance's Zhang also stressed to the HR department that employees' compensation packages should be "top of the market." He thinks that maintaining competitive compensation will increase the need for companies to deploy and perform well. This practice is similar to Netflix's in hiring the best people with top compensation.

Therefore, at ByteDance some outstanding employees have the chance to receive year-end bonuses equal to 100 months of their salaries, and good performers are likely to receive year-end bonuses equal to 20 months of their wages. The bonuses given by these high-tech firms are comparable with those given by the top investment banks on Wall Street.

In addition to monetary incentives, companies also offered gifts at traditional Chinese festivals, such as mooncakes for the Mid-Autumn Festival and rice dumplings for the Dragon Boat festival. These are common practices that exist in both public and private sectors in China; however, tech companies often employ unique designs for gift boxes, which could often become a social media sensation as employees of different companies compare examples of the mooncakes they have received.

Tech Firms' Preference for Young People

Since the beginning of tech in China, large companies have shown a preference for new graduates wherever possible. An early employee of Alibaba told us that there are multiple reasons:

- New graduates are easier to train—as much of what tech companies do is new, experience is often less relevant than being able to learn quickly.
- New graduates tend to be more motivated, compared to their more experienced peers.

- New graduates are easier to be mold to the company's values.
- New graduates are relatively easier to incentivize. You can get them to produce three times the results by paying them twice the market salary.

The same logic is also shared by Huawei to approach fresh graduates every year. In addition, the hiring philosophy of Wall Street banks or the consulting industry is not much different from this. Young talent is brought into the company on a regular basis and subsequently put through a demanding work environment with handsome pay; those proven capable are quickly promoted, while the others get filtered out.

There is recent evidence that many companies started to retrench "older" employees if they cannot contribute further to the company and fail to get promoted by a certain age. What is brutal is that the cutoff age for "older" is typically set around 35. This shows the stiff competition for talent that top tech firms are facing, which is very different from that in American firms like Facebook and Google.

In ByteDance, 11% of its employees were born after 1995 (younger than 25 years old); 52% were born between 1990 and 1995 (around 26 to 31 years old); and only 3% are above 40 years old. Interestingly, the gender ratio between male and female is 57:43, compared to Google's 68:32 and Facebook's 63:37.

Pushback Starts to Emerge

For a long time, working in big tech has brought promising prospects, especially for young people—their salaries are higher than peers' in nontech, they get rewarded handsomely with stock options, and the names of the company they work for make their families proud.

However, no company is able to sustain high growth indefinitely—while US tech firms such as Apple and Microsoft succeeded in catching new growth pillars, many large tech companies are struggling with intensive competition on all fronts. Baidu is a good example of this; so was Didi, the dominant ride-hailing giant that tried to launch into food delivery, financial services, and community group buying but was pushed back in each of these sectors because of competition from strong incumbents.

High growth masks a lot of problems for companies and countries as well. When that growth slows down, people's attention and motivation will change. Parallel to this is the massive uplift in living standards for the young people, which alters expectations. The hardship and suffering, preached by leaders who are of an earlier generation, start to lose their resonance.

The famous 996, which means working 9 a.m.–9 p.m. six days a week, lost its shine from what people would be proud of to something that young tech employees resent. In 2019, Jack Ma said, "Being able to do 996 is a reward; many wanted to do it but do not have the chance." He probably meant it, and if he had made that statement 10 years earlier, people would have seen it as words of wisdom and true encouragement; but in 2019 he met almost-unanimous backlash from media and social media.

Like young people in most places, those working in tech companies in China generally resent the long working hours and vent their frustration over social media. However, unless the competitive landscape eases, it is hard to foresee change.

This is when the government decided to intervene, setting boundaries of competition but also forcing the tech players to ease overtime for employees as part of the bigger strategy to halt plunging birth rates and avoid a demographic crisis. As a result, ByteDance and Tencent replaced 996 with 1075 (working 10 a.m.–7 p.m., five days a week) and 965 (working 9 a.m.–6 p.m., five days a week), respectively, with prior permission needed for anyone to work after 7 p.m. on a weekday.

Other environmental factors are driving changes, too: for example, when Beijing suffered persistent air pollution in the past few years, considerable tech talent we know personally moved to Hangzhou, where air quality was better. The move benefited Alibaba and the broader ecommerce/fintech ecosystem based in the city. Another example is when Shenzhen's living costs (especially property prices) became unbearable, many young people who had graduated from universities in Wuhan (yes, where COVID-19 was first detected) flew back to Wuhan, giving rise to a local ecosystem of tech companies.

SENIOR RANKS: THOSE WHO MATTER THE MOST

While companies are going all out to bring in talent at all levels, every leader agrees that the senior ranks, including their lieutenants, are more important than others.

"People at the middle level and below, while important, will not differentiate us from our competitors," the CEO of a $100+ billion company once shared, bluntly. "Our competitors have access to the same talent, and their capabilities will be very similar to ours."

"The top-level people, however, are where we can build a winning edge," he added.

The senior ranks of these companies are typically comprised of three types of people:

1. Those who started the company with the founders, or joined very early on
2. Experienced professional executives who joined the company at the growth stage
3. Senior managers who were promoted through the ranks

Experienced Senior Hires

Among the three types from the preceding list, the odd one out is the second: experienced executives hired from outside. Professional managers (职业经理人) often carry a negative connotation in China's tech sector. People see them as, while experienced and capable, not having the same level of commitment, risk taking, and resilience as entrepreneurs.

On the one hand, firms need experienced senior professionals to join in, as we discussed earlier in the chapter. On the other hand, senior hires often face issues with expectations and culture fit—failure is quite common, and companies need to be more careful in selection and expectation management.

In addition, a practical issue is how to put the right people at the right senior positions when the company grows and the demand for senior managers shifts.

This is crucial, as at junior levels, as long as the strategy, processes, and systems are set right, execution will not see much difference; however, a more or less capable or suitable executive at the senior level can make a lot of difference.

In addition to identifying, onboarding, and incentivizing such capable executives hired from outside, another key task of the leaders is to properly manage the relationship between these managers and the original founding team members.

Veterans and Senior Managers Promoted Through the Ranks

For the other two groups of senior ranks, *one issue is how to manage veterans who become complacent or lose motivation.* This can be a tricky problem, especially after an IPO when veterans become personally rich. Emperor Taizong, who founded the powerful Tang dynasty in the seventh century, once said, "It is easy to conquer the territories, but much harder to defend it," pointing exactly to this phenomenon.

Chinese internet firms have a special name—"old white rabbit." Old means that they have been in the company for a long time; white rabbit means that they look nice, harmless, but cannot compete or lack motivation to continuously create value to the company.

How to deal with these old white rabbits? One option is to kick them out. For instance, in February 2019, Richard Liu of JD announced the layoff of 10% of its management at the vice president level and above, including some of those who joined in early days.

The other option is more lenient, keeping these veterans as "culture ambassadors" in the organization, a symbolic position without real decision power (therefore not blocking or slowing down any real business). For example, some of Alibaba's veterans show up in various culture training activities for new hires.

The other potential issue of veterans is the power they gain over time, which may lead to corruption. A company's culture, value, integrity, and even survival will be jeopardized by corruption.

Many Chinese internet entrepreneurs, much like Chairman Mao Zedong, look to history for inspiration on how to prevent corruption and keep key officials accountable, a key theme throughout the dozens of imperial dynasties across Chinese history. Swings from brutal crackdowns to total chaos were common. Emperor Hongwu, who founded the Ming Dynasty in the fourteenth century, executed nearly half of his top lieutenants for corruption—his descendants, however, failed to control the rampant corruption, a key reason for Ming's downfall facing the military-disciplined Manchus in the seventeenth century.

The same is happening with Chinese tech companies. When executives hold power over suppliers, partners, and other stakeholders, corruption arises. Anticorruption investigations often lead to high-profile sackings and even arrests.

CORRUPTION IN CHINESE INTERNET FIRMS

On April 21, 2020, Baidu's Professional Ethics Committee reported a case of corruption: Fang Wei, a former vice president suspected of corruption following an internal investigation, was handed over to the police.

It was not the first time that Baidu reported internal corruption but rather it is part of a series of antigraft campaigns. On July 31, 2019, Baidu reported 12 cases of internal corruption, including bribery and infringement of trade secrets, as well as disciplinary violations such as false expense claims and false reporting of bonuses. In May 2015, Baidu fired eight executives in criminal investigations, including senior ones in the sales and marketing departments.

Like Baidu, Alibaba, Tencent, Huawei, and some of the most prominent tech firms in China have internal investigations departments. ▪

Jack Ma of Alibaba famously adopted a zero-tolerance stance toward any corruption or violation of trust cases. In 2011, after David Wei, then-CEO of Alibaba.com, was pushed out for endemic violation of trust under his purview, Ma sent a note to all Alibaba employees and clients, saying that the Alibaba team needed to "face the reality, shoulder the responsibility and have the courage to endure the pain of scrapping the bone to rid the toxin."

He also realized that, maybe specifically in China, just having the standard rules and processes alone were not enough. Alibaba needed more than that, not only to stamp out corruption, but also to ensure that people at Alibaba are aligned to keep the organization running for the long term.

With the aforementioned Savio Kwan, Lucy Peng, cofounder and later chairwoman of Ant Group, and a few other key executives, Jack Ma embarked on an iterative journey to design Alibaba's value and HR systems; many of the lessons learned and schemes designed along the way still hold much relevance to this day.

ALIBABA'S PEOPLE AND HR SYSTEM

A company's people policy is embedded in its value system, the foundation of any long-lasting organization. Jack Ma said that "Alibaba is a unique company because of our culture and value which make us who we are, but not someone else. It allows us to make the right decision at the most critical and critical moment."

The Evolution of Core Values

Ma and his management team created their first version of the core value system in 2001, which consisted of nine key pillars: passion, innovation, learning and teaching, openness, simplicity, teamwork, focus, quality, and service and respect. Ma, who was a big fan of martial arts novels, called these pillars "nine swords."

However, how to build these values into the team, and make them useful for the team, became a challenge. Most of the early employees of Alibaba were very pragmatic with a fighting spirit; a common thought at the time was that the "values" were too fluffy for a company that was still fighting for its survival and were, basically, a waste of time.

Ma took learnings from Mao's early movements to "use a value system to align the thinking, and use the aligned thinking to guide actions, and form a coherent force." The management embedded the core value system into its compulsory training, especially for its business development teams—in fact, the value system, taught by Ma, Kwan, and Peng, consisted of 60% of the training time, while the other 40% was on actual business skills.

Such practice was novel for any company in China at such an early stage, but it bore fruit. Those who did not believe in the value system quickly decided to leave, and those who remained worked coherently together.

The "nine swords" later evolved into "six divine swords":

1. Customer first
2. Teamwork
3. Embrace change
4. Integrity
5. Passion
6. Commitment

"Customer first" guides employees to set the right priorities when facing conflicts of interest; "teamwork" and "embrace change" are principles for teams interacting with external environments; while "integrity," "passion," and "commitment" are basic qualities employees need to have.

In 2019, at the 20-year anniversary of the group's founding, Alibaba further refreshed its value system again, and the six new principles are:

1. Customers first, employees second, shareholders third
2. Trust makes everything simple
3. Change is the only constant
4. Today's best performance is tomorrow's baseline
5. If not now, when? If not me, who?
6. Live seriously, work happily

Daniel Zhang, who had succeeded Jack Ma as Alibaba Group CEO by then, said that the key objective of the new principles was to "find people who can be with us on the journey for the next 5, 10, and 20 years."

Evaluation and Assessment

Throughout the reign of Jack Ma, Alibaba attached great importance to the values of its employees. For its quarterly assessment of its employees, values account for 50% of its total weight, while results account for the other 50%.

Interestingly, in its early days, Alibaba divided its employees into five categories based on the assessment, using very bland language:

1. Dogs (poor results, poor values)
2. Wild dogs (good results, but poor values)
3. Little white rabbit (poor results, good values)
4. Cow (average results and values)
5. Stars (good results, and good values)

It was relatively easy to deal with all the categories other than "wild dogs." However, in early years, after the group ruthlessly fired employees in that category, the whole organization realized that management was serious.

Training for Managers

Alibaba's managers are systematically trained on how to deal with people. They are told to "Dream big, start small, and learn fast."

Fundamentally, managers need to understand that the team can only succeed when the team members' personal goals are aligned with the team goals. In order to achieve that, they need to uncover each team member's personal goals, and also continuously explain the team's goals to the employees. It is a recurring four-step process:

1. Deeply understand the team's goals.
2. Form the team's "dream."
3. Uncover individuals' personal dreams (can be material, growth, or responsibilities).
4. Align the team and personal goals and dreams, and continuously sell.

"Continuously sell" is a key reminder that managers need to use all the channels—verbal, emails, recruitment, and so on—to emphasize the key messages internally and externally. For the employees, the best timing to sell includes onboarding, their (quarterly or annual) review, encountering difficulties, or achieving key successes.

Managers are also reminded that "Employees are an asset lent to you by the company. It is your responsibility to add value to them." Therefore, managers need to constantly ask themselves, "Has the value of my team members increased or decreased while they are with my team?" "Are they merely completing tasks or continuously growing professionally?"

Taking care of team members' personal emotions is also a key responsibility; otherwise, "when a setback comes, their first thought would be leaving."

Another important aspect of the managers' training is how to deal with people who are not performing or are not suitable for the role.

This starts with hiring. Managers are asked to assess the candidates on six criteria before making a decision:

Yes:
1. This person will bring long-term value to the company.
2. This person will enhance the strength of the team.
3. We will suffer if this person goes to a competitor.

No:
1. This person should be able to do the job well. We have a shortage of people.
2. I'm a little hesitant, but I don't think this person has got anything wrong.
3. Business pressure is very high, and it is good to have someone covering the job.

On the firing side, managers are asked to reflect first on the question "Have I helped the person adequately?" before making a decision. They also need to remember some simple mantras:

- "While the heart has to be benevolent, the knife should be fast."
- "Letting the person go is for their own benefit."
- "Happy stay, happy go!"
- "Remove the weeds on time."

A horrifying situation would be "those who should leave don't, those who should stay, leave."

Also, oftentimes a person should leave their current role not only because they are a bad performer; in situations where they are outperforming the role they are in, or have exhausted their growth potential in the role, they should be given more responsibilities, rotate to another role, or be entrusted to train the team.

All these concepts are distilled into very simple principles and mantras, and used extensively in internal training. Also, as they preached, training, just like managers selling the dream, needs to be continuous and repeated.

The Commissars

Lucy Peng, who acted as the first chief people officer, also instituted a position called HR generalists. They took their model from how the Communists led their armed forces: in each unit, big or small, there was always a commander and political commissar. What does the commissar do? Ensure the political and doctrine education in the unit they are assigned to.

The thought behind Alibaba was that while the CEOs and business leaders of each unit are good at getting business results, they might not necessarily be suited to also institute the values. This is because they come from different backgrounds, have different management styles, and themselves might not be very good at dealing with people and value issues.

Therefore, the HR generalists take on the role of educating the workforce of the value and higher purpose of the organization.

Lucy Peng once mentioned, "When the CEO does not know what is right for the organization, the CPO should have the guts to nudge and remind the CEO."

The organization, therefore, places high requirements on the CPO as well. They need to understand the business, because otherwise their nudging and suggestions might push the team in the wrong direction.

However, a challenge that emerged later on was how to prevent the HR generalists system under the CPO from becoming stagnant, as they were not really responsible for business results. We can see this happening in some of Alibaba's overseas units.

The Partnership

Such schemes ensure that values are at least understood and respected by every employee. However, over the long term, as any incentive system designed around metrics, they become easy to game. Also, as the competitive and talent landscape keeps evolving, the way values are communicated, or the values themselves, need to evolve as well, presenting great challenges to these organizations.

To uphold the integrity of Alibaba Group, and also to ensure longevity, Jack Ma has created a partnership scheme. The system, which is very different from other types of modern corporate governance, is quite controversial. We will see whether it works out or not after two or three leadership successions.

PEOPLE ISSUES IN OVERSEAS EXPANSION

While leaders' commitment to overseas markets is critical, any expansion initiative has to be implemented by people. Despite that, Chinese tech companies recognized the importance of people and devoted huge investments in

ALIBABA PARTNERSHIP

Alibaba partnership was founded in 2010. Unlike the general partnership system, Alibaba's partnership system is unique in that the partners are not responsible for operating profits and losses. Also, the most important purpose of Alibaba's partnership system is that it is partners who have the right to nominate a majority of the directors on the board rather than allocating board seats based on their shareholdings.

Partner candidates must meet several criteria: (1) they have integrity; (2) they have more than five years' work experience in Alibaba; (3) they have made a significant contribution to Alibaba Group; (4) they are a "culture carrier" who is consistent with the mission, vision, and value of Alibaba.

The Alibaba partnership committee is required to be made up of at least five but no more than seven partners, including partnership committee continuity members. The committee is currently composed of Jack Ma, Joe Tsai, Daniel Zhang, Lucy Peng, Eric Jing, and Jian Wang.[1] ▪

terms of time and effort to identify, select, motivate, and retain their people at different ranks; going overseas presents another set of complexity and a big challenge on the people front. What are the key hurdles and how does the people gap get filled?

Cultural Context: The First Mindset Shift

The people issue is a two-sided problem. On the one hand, in our interviews we often hear overseas employees, including senior managers, complaining about the trust, communications, and cultural issues within the Chinese companies they work for. On the other hand, these tech firms also feel frustrated.

Firms need to identify the right people to hold key responsibilities in overseas expansion, and when such people are hired in the local market, the culture, including ways of communicating and the context, can be quite different.

For example, mobile payment in China has been prevalent for a long time with a duopoly of Alibaba and WeChat, to the point that when building a product, Chinese managers do not think payment is a problem—but that is not the case in many markets, and the lack of this fundamental understanding often makes communications ineffective.

Often, not only the leaders, but also other teams or executives in China would need to face the overseas managers, and misunderstandings and frustrations can happen in areas and places where leaders least expect. This can cripple trust and operational effectiveness.

What makes things complicated is that Chinese culture has very high context, that is, many messages are not explicitly communicated but are embedded in the common context. The fact that many of the markets Chinese tech companies operate in, such as Indonesia, Thailand, and Saudi Arabia, are also high context but with different context, makes communication even more difficult.

In this regard, American companies such as Amazon have big advantages, because the American culture is inherently much lower context (direct), and in many countries you have enough elites who have received Western education and therefore are well versed in communicating with Americans.

For example, Amazon invested billions of dollars in launching and expanding its Indian operations. Amit Agarwal, global senior vice president and head of India, was originally from India and had spent 14 years with Amazon before heading to Bangalore to launch the India business. Many of the senior executives in Amazon's Indian operations were Indians who had worked with Amazon. Additionally, because Amazon functions in the English language, it is no problem for other Indian executives to fit in. Similarly, Amazon's Middle East operations are able to fill its ranks with Western-educated executives who are well-versed in both English and Arabic.

In comparison, when Alibaba started looking at India more seriously, it identified less than a handful of Indian executives who could speak Chinese and coordinate effectively with teams back in China.

Recognizing this, in our opinion, is the first mindset shift that Chinese leaders and managers need in order to function effectively with people in expansion markets.

How to Fill the People Gap?

Who are the best people to hold key positions and make key decisions in the market, acting not only as eyes and ears, but also arms, legs, and part of the nervous system for the leadership in China? The companies we studied, in their respective journeys, have tapped into different options.

Tap into Old Hands at Other Chinese Firms

Executives from Huawei and ZTE, two telecom equipment companies that are long-time pioneers in overseas operations, became top targets for headhunting when Chinese tech firms ventured overseas. For a while, Alibaba Cloud built almost its entire overseas sales force with former Huawei executives.

In areas like government relations and in markets like Indonesia and Brazil, some also hired executives previously working for infrastructure and/or state-owned Chinese firms.

Even without hiring these people, Chinese interest companies would often at least consult the Chinese business community in the destination country, in an attempt to understand the local market more from a Chinese perspective.

While some of these executives work out brilliantly, across the board the big challenge of culture fit persists. Most of the previous generations of Chinese firms overseas, like their Japanese and Korean counterparts, are focused on selling products or solutions to businesses. The executives are typically very experienced in B2B sales, distribution channel building, and government relations.

While these skills are certainly very important in their respective countries, they are often different from the core skills of consumer tech companies, including product, branding, user operations, and crisis management.

Another challenge we have seen is pathway dependency, meaning experiences from previous successes dominate the way decisions are made and actions taken. Senior executives from a different industry might not be as agile in tech businesses. We have seen cases in Indonesia as well as in Saudi Arabia where business models that should focus on consumer user acquisition and customer experience instead are devoted—with outsized attention—to government relations and partnership building, leading the expansion efforts in the wrong direction.

As we said before, overseas venturing is the second entrepreneurs' journey. While assessing, entrusting, and helping the hires from more traditional Chinese companies already in the market, leaders will need to be aware of these potential mismatches, and address them early.

Hiring Local Talent in Foreign Markets

A more aggressive, and perceived to be faster, approach is identifying and hiring the best local managers in each foreign market. We have the success of Xiaomi mentioned in Chapter 4, Leadership, as a good reference and benchmark, right?

JD.com trusted the Thai management for its fintech joint venture with Central Group in Thailand; JollyChic hired someone from Souq.com (later acquired by Amazon) to be the general manager in the Middle East; Didi hired a group of ex Uber, Easy, and Cabify managers when it expanded to Latin America.

However, this has been proven challenging in many cases. JD's managers in Beijing complain about Thai managers being slow; JollyChic's Chinese managers bypass the Middle East general manager to coordinate directly with other departments; and many of Didi's local executives in Brazil left quickly.

Even in Lazada, the original approach of having some Alibaba executives work alongside the existing management (prior to acquisition in 2016) did not work out. As the company was losing ground to the competition, a suite of Chinese managers was parachuted in to replace a large part of the existing management in 2018.

Many have simply attributed these problems to the heavy-handed or top-down management approach of Chinese companies. However, this is a rather simplistic assessment of a much more nuanced picture.

A key issue here is the aforementioned culture context, that makes communications and expectations alignment difficult. Western companies spent years and, in many traditional industries, decades, to figure out markets and culture codes (a still evolving process). For many Chinese tech companies, their overseas journeys are young, and the lack of adequate culture awareness and training means that alignment is often hard for both sides.

A related issue is language. Many managers from Chinese companies, especially of the generation of Alibaba and Tencent, often speak English poorly or not at all. The difficulties in language add to the possibilities of misunderstanding as well as frustration, as people on both sides are stuck in situations where multiple clarifications need to be made.

Second, in many tech sectors, the evolution in China is far more sophisticated than that in many other markets. Local executives outside China, who have not been immersed in such development, tend to find it hard to appreciate or apply the learnings from there. This is like adding another layer of context to the already complicated communications barrier.

For example, Chinese internet managers tend to understand consumer traffic, growth, and user operations very deeply, while it is harder for their counterparts in destination markets to be on the same page. We have seen cross-border teams argue about what "marketing" really means in many cases (yes, even that and the fundamental methodology could be different).

Third, the expectations of speed are different. In China, things happen fast, whereas in many developing countries, tech companies operate at a relatively slower pace. This expectation is often not communicated or aligned well, causing premature decisions to be made and strategies to shift. We mentioned in Chapter 4, Leadership, that Kuaishou decides to increase or decrease investments on destination markets because of how the market responds to user acquisition campaigns, not the underlying reasons behind such responses.

There are many other examples of where a decision is set, and in three months no results are achieved, and therefore the decision reversed—however, communications along the process failed to make people understand that it simply takes at least one or two months for the products to be updated, and another couple of weeks for users to adapt to the new product; therefore, seeing results in three months as would be expected in China is not realistic in many markets. The simple concept of time carries different meanings in China than it does overseas.

The fourth issue is the culture, as we mentioned at the beginning of this section. Alibaba's value system is very unfamiliar with what especially the Western-educated executives went through in their experience. Documents and training materials are simply translated (often poorly) from Chinese. Executives hired from outside often find it hard to resonate with the culture and get dejected easily.

QR CODE

In 2018, one of us was personally involved in a consumer tech project that has operations in Jakarta, Indonesia, and tech development in Hangzhou, China.

The Indonesian team explained that fraud for e-discount vouchers could be a big problem—the Chinese tech team immediately designed a complex scheme to prevent such loopholes. Led by a former Ant Group fraud specialist, the system was designed, developed, and went live within just two weeks.

The only problem was that nobody in Indonesia, including operations and customers, knew how to get around it. As a result, nobody could use the new release of the product.

A few rounds of talks to resolve this issue failed to achieve anything. It was only when we flew the fraud specialist to Jakarta, where he observed local fraud patterns for a week, that we found a solution that was secure and usable.

Upon leaving, he said, "I did not expect the frauds here to be so simple and easy to deal with." What he had originally designed would, according to him, capture 90% of the most sophisticated fraudsters in China. Such fraud patterns only started appearing in Indonesia one and a half years later, when some of such fraudsters in China moved over because it became more difficult for them in China.

More such issues and their analyses are discussed in Chapter 7, Product. ■

Many companies have realized this problem—and we think their approach of sending their foreign managers to headquarters for mandatory training or hiring management trainees globally to work in China first is a good step. This is something American companies have been doing for years. The devil here, again, is in the details (i.e., how to execute such programs effectively).

Sending Expats from Headquarters

Like American and European companies, or Japanese and Korean firms, for the past few decades Chinese tech companies also sent expats. Expats solve the experience and expertise issue, especially when speed is needed. They know the company history, products, and the organization's culture, and they can also translate local reality for headquarters and top-level decisions to local operations.

However, we all know the drawbacks of expats. Many of them are not meant to be stationed in the destination market for a long time, and therefore lack the commitment to understand the market deeply or undertake long-term initiatives before they are transferred out.

This issue is less a problem for companies like Huawei, which rely on selling B2B products. For companies that build consumer-facing ecosystems, the lack of long-term thinking and the ability to balance that with short-term goals can be a serious issue.

To balance that, systems need to be built much like institutions in democracies that balance out the short-termism of elected leaders.

Systematic Capacity Building—Management Trainee Programs

Recognizing that this is a long-term commitment, some companies start to systematically build an international talent base internally. For instance, some tech giants, such as Alibaba and JD.com, have launched global management trainee programs, which seek young talent all over the world.

While this approach makes a lot of sense, the execution of it is often difficult. First, the internet is the space where things change rapidly, as mentioned many times throughout this book. When organizations' priorities shift frequently to respond to market demands, it is very hard for a long-term trainee program to show results. As the organization changes its priorities, many trainees will be disoriented, resulting in suboptimal results or exodus. This has happened in trainee programs of more traditional companies as well, but in tech the churn is often accelerated.

EXPATRIATES FROM ALIBABA IN LAZADA

After acquiring control of Lazada, Alibaba successively sent a number of senior executives to Lazada to work in the Southeast Asian market. However, the Hangzhou-based executives' performance in Southeast Asian markets has been less than satisfactory. The result has been a series of senior changes at Lazada in recent years.

Outsiders often blame the Lazada executives' lack of cultural understanding and heavy-handed approach in managing local teams for this. However, while this might apply to a few who just wanted to achieve some quick results to get a promotion back home, we see many more from this cohort who genuinely wanted to make a change and grow the business. Remember, these are carefully selected Alibaba executives who have had good track records.

What happened, then? Those who want to make a change are often stuck communicating and coordinating with different departments in Lazada's headquarters in Singapore, Alibaba's home base of Hangzhou, and sometimes other Alibaba-affiliated teams such as Ant Group and Cainiao for payment and logistics, respectively. As many people who are not in the market are in the decision-making process, things are naturally slow. They need to absorb information first, and without firsthand experience it is hard for them to develop good business sense and judgment. This, coupled with the expectation for speed, led to cycles of flipflops in direction. ▪

Second, many companies do not have established experience in international markets to give this young talent sufficient training and exposure. Quite often, it just became a program where international young talent is learning the company's operations per se, detached from their markets.

ALIBABA GLOBAL LEADERSHIP ACADEMY

In 2016, Alibaba launched the Alibaba Global Leadership Academy (AGLA), which aims to create "an innovative, 10-year program, to enhance its global workforce capabilities."[2] It is a Hangzhou-based program that recruits 102 candidates from all over the world and provides rotation training for 16 months with the purpose of supporting Alibaba's international growth. In addition, the participants can learn the ecommerce business culture in China and Chinese culture, history, and arts.

After finishing this rotation program, the participants will be sent overseas to lead the local team to expand international business. ▪

JD'S INTERNATIONAL MANAGEMENT TALENT (IMT) PROGRAM

In 2013, JD.com launched the International Management Talent (IMT) program, which is a three-year tailor-made development plan aiming to cultivate leaders with a global perspective.

Accordingly, candidates must graduate from the MBA program of top global business schools with ambitious, solid, and grand career visions. Then they have the chance to apply for full-time positions. For example, the first batch of candidates is all fresh MBAs from prestigious schools such as Wharton, Sloan, INSEAD, and London Business School. Richard Liu, the founder of JD.com, hopes these candidates will become the next generation of leaders.

Once admitted, candidates will undergo a 10-month rotation in each diversified business unit at JD.com with the supervision of C-suite executives and VPs. Candidates will gain systematic business-related knowledge from each rotation and their mentors, including purchasing and sales across different categories, warehousing and delivery, importing and exporting, customer service, big data and cloud computing, consumer finance, insurance, payment solutions, supply chain finance, and so on. ◾

In Alibaba, as well as in JD and many other more traditional MNCs, trained future leaders do not stay with the organization once their rotation program finishes. This has a lot to do with the execution of the program, as well as the expectation of management, as mentioned earlier.

Here, we would like to highlight that building an international talent base from scratch is *not* easy. Companies should have the appropriate expectations for such programs, and the leaders should set long-term commitments and align their teams toward them, just as they do in their domestic hiring and training strategies.

 CRACKING THE CODE

Despite the difficulty of filling the people gap in overseas expansion, we found that a few companies might have cracked the code. For instance, TikTok in Brazil, according to a friend who is familiar with the development, initially tried to hire local senior executives, or senior Huawei or China state-owned enterprise (SOE) executives who have been pioneering and living in Brazil. They found that most of these people were not compatible because of both culture, and, ironically, a generation gap—as TikTok's targeted audience is teenagers.

The company eventually resorted to hiring some young Chinese graduates who spoke Portuguese. They are young, fit in both cultures, and are also
career- and aspiration-wise attached to Brazil. In addition, they can relate to
the young, aspiring Brazilians TikTok tried to bring on board. They proved an
important force to help TikTok take root in Brazil and continue its growth.

While companies can adopt a combination of the approaches highlighted
in the section "People Issues in Overseas Expansion" and understand the challenges and difficulties of each, we have summarized some general pointers here:

- What matters is the fit between people's expertise and the company's product and targeted market. Such an alignment can best utilize the people's
  knowledge, skills, social capital, and networking in overseas expansion.
  If you hire an executive with a strong background in the B2B space to
  manage a B2C business, you need to pay some extra attention to the process and progress.
- Expansion requires people's commitment in the overseas market, no matter whether they are externally hired and local, or internal expatriates
  from the headquarters. Without strong commitment, your people will not
  have a deep understanding of the local market and will not see the future,
  and therefore won't be inspired to conquer the market, especially when difficulties arise. The commitment of people also needs corresponding effort
  and long-term vision from leadership, which we discussed in Chapter 4.
- The people managing overseas expansion need to be culturally sensitive.
  They need to understand the organizational/industry culture, Chinese
  culture, and the culture in each market they are in charge of. Hiring overseas Chinese managers in each market seems to be a potential solution,
  given the widespread presence of ethnic Chinese in various markets, especially in Southeast Asia and Latin America. However, most of the local Chinese grew up and were immersed in very different contemporary contexts
  compared to their counterparts in China—misalignments of goals, expectations, and culture understanding can be quite common here as well.
- Companies need to be aware of and address the motivation gap. While
  extrinsic incentives such as money are universal motivators, they work
  differently in different cultures, social economic standing, and personal
  aspirations. We have often heard Chinese managers complain that local
  employees in certain countries will not work extra hours even if paid three
  times their usual rate.

Fundamentally, forcing incentives, such as higher pay with longer working hours, do not resonate with many local hires, and sometimes can be

counterproductive. Indeed, Glassdoor data shows that the overall rating from the 8,000+ employee reviews of Huawei is only 3.7 out of 5. Baidu achieves 3.9. By contrast, Cisco Systems has a score of 4.3, and Google is 4.5.

Recognizing that money cannot generate the same degree of incentive power, and that people's aspirations can be different, Chinese companies working outside China need to actively identify what strings to pull to build the right employer branding, attract the right talent, and align the goals of the people in foreign markets with the companies' success in those markets.

In that regard, we feel there is actually a lot they can learn from non-tech multinational companies that have been operating successfully in those markets for decades.

Ultimately, Chinese leaders are fast learners with deep reflections.

QUESTIONS TO ASK

Similar to Chapter 4, we would like to ask a few questions for leaders to consider to conclude this chapter. These could be the checklist on people-related issues when you aim to enter overseas markets.

- Who do you have to lead your expansion efforts in the market? How do you ensure that you deploy the best people possible for the success of such expansion?
- If that person underperforms, how do you find the reason, and what would you do?
- Do your people in other parts of the company who need to support the expansion have an adequate understanding of the market, a willingness to support the expansion, and, if necessary, a willingness to relocate?
- Do your company's core values and HR system appeal to the right talent for the market(s)? If not, what would you do?
- Do you understand the people and culture of the market well enough that you can communicate with people there and interpret well what they tell you?
- How do you give your people their objective assessments? What would incentivize the key people you plan to hire, and keep them motivated and retained in those markets? If it is not only money, what else will work, and what is their relative importance?
- Do you know well enough the experiences of other Chinese/international companies that have cracked the people code in that market?

NOTES

1. Alibaba Group, "Corporate Governance," https://www.alibabagroup.com/en/ir/governance_9.
2. Alibaba Group, "Alibaba Group's Innovative Recruiting Program Supports International Growth," press release, October 24, 2016, https://www.alibabagroup.com/en/news/article?news=p161024.

CHAPTER SIX

Organization

Managers have the incentives to push the company to set more and more explicit rules and processes. This push is like gravity, which keeps pulling the organization towards more complexity. During this process, we must actively resist the gravity, and keep stretching the organization.

—*Liang Rubo, cofounder of ByteDance and CEO since 2021*

BYTEDANCE'S NEW CEO, LIANG Rubo, was in a discussion with a manager in the R&D team, who said that a really promising engineer had been poached by another company for twice the salary.

"If you really think this person is promising and valuable," Liang asked, "why did you not match the incentives, instead letting a competitor take him away?"

It turned out, the manager had tried, but in the HR system the maximum salary he could secure was not enough to match the competitor's offer.

As Zhang Yiming decided to free himself from the day-to-day operations of ByteDance to focus on long-term strategic decisions, including international markets, cofounder Liang Rubo took over. Liang, who was Zhang's university classmate, now had to lead the 100,000-people organization. He needed to deal with challenges like this. On the surface, it is a people-related issue. But fundamentally it's an organizational issue.

Liang was taken aback by this episode. The manager had been with the company for a number of years, yet he was not able to act in the best interest of the company because of the system's limitations.

Liang and Zhang, both avid followers of Netflix's "context, not control" culture and users of OKRs (objectives and key results) rather than KPIs (key performance indicators), realized that in a large organization, more needs to be done to ensure that the implementation matches the theory.

It is natural for large but still growing organizations to increase their arsenal of rules, and to make those rules more explicit. Otherwise each decision will be escalated to the upper level and need to be studied, deliberated, and aligned—not an efficient process. However, with more rules, the organization, while becoming more efficient in decision making, might become more bureaucratic, rigid, and less adaptable to change.

In the fast-changing tech world, such rigidity could become fatal very quickly. Therefore, how to keep their organizations agile and adaptable is always at the top of the agenda of Chinese tech leaders.

The challenge is universal. Globally we have seen many stories of great companies declining into irrelevance, often due to the inflexibility of a large, once successful, organization. Much business literature, including case studies in business schools, has been written, including the famous classic book *Who Says Elephants Can't Dance?* by former IBM CEO Louis Gerstner.

The conventional argument goes like this: as a company becomes successful, its organization becomes rigid, executives complacent, and change is impossible. Even though the top leadership often knows acutely where and what changes are needed, implementing them against the inertia and complexity of the organization is extremely difficult.

What types of structure, rules, and processes need to be in place? How do you navigate an organization of 100,000 people with multiple departments and business units (like many large Chinese techs now) through a constantly changing landscape?

Chinese leaders can also draw lessons from history. Many ancient Chinese dynasties, including the Han, Tang, and not-so-distant Qing, fell into terminal decline very soon after reaching a golden period. Emperors and reformist mandarins throughout these dynasties have made various efforts to arrest the decline, and they almost always failed.

To avoid such a fate, especially in an environment of hypercompetition where everything (including decline) goes much faster, almost all the large tech companies in China put "organization" as a top agenda item. The organizational practices of Alibaba, Huawei, and Ping'an almost dominate the business sections of bookstores in China.

However, when these companies adapt their successful organizational practices to overseas ventures, another set of challenges arises, making the organizational process a constantly evolving and often frustrating one. The lessons we can draw from this are explored in this chapter.

THE KEY ORGANIZATIONAL ISSUES

Needless to say, an organization is an organized group of people with a particular purpose. It contains both the static perspective of a company's structure and the dynamic perspective of a process in which decisions are made, information is communicated, and resources are allocated in a firm (Figure 6.1).

The central question is how to form the right organization to deliver the right product to meet the customers' needs and adapt to changes of the environment.

FIGURE 6.1   Key issues in an organization

DIVISION AND INTEGRATION

The fundamental issues of organization to consider are "division of labor/tasks" and "integration of efforts," as professor Phanish Puranam suggests in his book *The Microstructure of Organizations*.[1]

In the division of labor, companies divide their activities into different tasks performed by different people and groups (atomization because of specialization-driven efficiency), which then need to coordinate to produce coherent output.

Large Chinese tech companies often do a good job with division of labor: establishment of business units, segmented positions in organizations, and detailed job descriptions. These concrete measures help firms divide their goals into concrete, measurable subgoals. But another problem arises simultaneously: how to integrate all the efforts into a coherent outcome and how to coordinate different divisions of labor to achieve the same goal.

What makes things more complicated is the potential agency problems of people at different levels of the organization—their interests may not be fully aligned with the organizational goal. That, coupled with the distorted information and fast-changing nature of the tech market, makes integration a nightmare.

To effectively manage task specialization and later integration (even at a suboptimal level), organizations need to address critical and intertwined issues, including the decision rights, information communication, and resource allocation. ▪

Decision Rights

A perennial challenge facing all firms and their bosses is how to allocate *decision rights*: *decentralization* or *centralization* (i.e., bottom up or top down)?

On the one hand, you want to give people on the ground flexibility; on the other hand, you would want the organizational resources to move more coherently, especially in a competitive environment. Besides, how do you make sure you always have the right people at different levels to consistently make the right decisions?

The often-touted flat management and empowerment of employees do not come from a preference, but from a necessity to survive and compete. Information needs to flow quickly and effectively, and resources allocated and readjusted in an equally speedy and effective way.

Ren Zhengfei, the founder of Huawei, once noted that people in the market should have the decision rights: "let the people who can hear the gunfire make the decisions." He was worried that Huawei would become a bloated, bureaucratic organization where executives are drowned in processes rather than market instincts.

However, "let the people who can hear the gunfire make the decisions" is a blanket statement that, while motivating, can't be implemented easily. We are not running a computer strategy game, where a decision input will give the (often instant) feedback of results. Many decisions in real life involve moving people and resources, and impact other parts of the organization.

For instance, consider this: you are the general manager of Southeast Asia for a Chinese cross-border ecommerce company in Hangzhou, and your Vietnamese colleagues want to readjust the supply chain to capture a new market segment, but doing so will cause a decline in your Thailand market, which is smaller but more profitable. Would the Vietnam country manager be able to make that decision? You? Or your superiors in Hangzhou?

It becomes not that straightforward.

Information Communication

When the communication of information is blocked or inefficient in an organization, decision makers will have incomplete information, or will not have the same understanding or interpretation of the information available. This is why organizations try to use different software packages to facilitate communication and mitigate potential problems.

In stable companies in mature industries, executives can do proper strategic planning based on the trove of information from market feedback, research firms, and established internal communication channels. In fast-growing companies in fast-growing sectors, you often need to make decisions long before structured information becomes available. Additionally, the information continuously updates and changes—new legislation, changing demand, competitors' attacks, new entry with fundraising, and so on.

Resource Allocation

Resources in an organization include money, materials, human capital, time, and other assets that are essential for firms and their business units to function effectively. The common issue is the limited pool of resources, no matter how big the organization is. The leader has to face the resource allocation decisions— what the priority is, how much to allocate, how fast the resources should be ready, and so on.

In an ideal situation, resources can flow smoothly like information, and support the decision rights. In reality, business units are competing with each other for resources, and estimates of resources required are often quickly smashed by changing market conditions.

To summarize, the organization is a structural issue that involves multiple elements. The rule of "structure follows strategy" proposed by Alfred Chandler's classic book *Strategy and Structure* (1962) is easy to understand, but hard to implement. Every case of failure to adapt a product to local realities can be attributed to failure in organization: it could be a mismatch between decision rights and resource allocation, blocked information flow or miscommunication in process, or a combination of these.

Chinese tech leaders are acutely aware of this—and they have taken many steps to proactively address the challenge.

 ## ORGANIZATION OF CHINESE TECH FIRMS

Successful Chinese tech firms typically have powerful and stable leadership who keep the decision rights. Information communication is channeled from top to bottom, and decisions about resource allocation are made at the headquarters. In general, organizations use a centralized structure at the top to ensure panorganization integration and that the overall goal is secured.

The question is to what extent, in terms of size and complexity, the organization can grow while people at the top can still have the mental capacity to make informed, optimized decisions.

To defy gravity and keep their organizations adaptable, leaders of large Chinese tech companies have made various efforts, often taking inspiration from history. We observe that Chinese internet firms have developed a number of tactics, including:

- Continuous organizational restructuring
- Rotation of executives
- Internal competition
- Internally developed productivity and collaboration tools
- Small, agile teams and the attempts of Zhongtai

Let's dive in with a little bit more detail.

Continuous Organizational Restructuring

Executives around the world are probably familiar with organizational restructuring—which happens once every few years to realign the structure and control costs, sometimes but not always accompanied by retrenchments.

What makes large Chinese internet firms unique is that they may launch small-scale and continuous organizational restructuring even if there is seemingly no need for major strategy changes.

For example, since Daniel Zhang took over as Alibaba Group's CEO in 2015, a major restructuring happens every year: portfolios of major business units are readjusted, executives reassigned, and new ventures incubated. Under these major restructurings, smaller adjustments happen at departmental levels, and even micro adjustments at team levels.

After announcing one of the changes, Zhang said: "Business environment and customer requirements are continuously changing, Alibaba is always growing with constant changes: from separating Tmall from Taobao to all-in mobile, every major organizational adjustment releases huge productivity."

He added: "Only when we are brave to change ourselves, and embrace the uncertain future, can we bring certainty in services and experiences to our consumers and customers. It is the best future proof for any company now."

These are only the major restructurings. Many friends who have worked at Alibaba have encountered various minor, departmental restructurings. The most prolific case we can personally verify is six times over a year. This is hardly healthy for a person's career development, but was deemed necessary to keep the organization adaptable.

Alibaba is not the only company that does frequent reorgs. Wang Xing, Meituan's founder, once told his team that "Meituan is always six months away from bankruptcy." The remark reflects Wang's sense of urgency on strategy, organization, and the competitive environment.

Meituan's business and organizational structure have been constantly adapting to the rapidly changing competitive environment. In 2015, it structured the company in five business groups, covering hotel, food delivery, dine-in, catering, and other platform businesses. In January 2017, the five business groups were collapsed into three to better reflect market demands and internal efficiency. By October 2018, a further two rounds of restructuring had taken place, regrouping existing businesses and adding new businesses.

Even in 2021, when ride-hailing giant Didi faced regulatory challenges and was forced to stop new user signups, Meituan quickly seized the opportunity, breaking its Smart Transportation Group and elevated ride-hailing business into a separate business unit reporting directly to the CEO. At that time, the Smart Transportation Group had been set up for barely a year.

Meituan's organizational restructuring, on the one hand, is to respond to the change in its strategy. On the other hand (perhaps more importantly), it's a top-down effort to inject urgency and energy to the organization against bureaucracy and complacency.

Rotation of Executives

There is a saying from imperial times: "The mountain is high, and the emperor far far away," describing that local officials in the vast empire would often govern without the central government in mind—they were too far away anyway.

To retain control, the first empire in China more than 2,000 years ago had already abandoned the feudal system. Instead, it created an alternative system where officials, including county magistrates and provincial governors, would have personally come from a different locality, and they would rotate. This system, if implemented well (a big if in imperial times), would prevent corruption as well as local vested interests having too much control.

The same system is adopted by the Chinese Communist country, and also by many tech firms.

Huawei and Ping'an, two firms revered for their organizational practices, are known for mandatory rotation of their executives.

The job rotation system (轮岗制) in Huawei is divided into business-based rotation and post-based rotation. For instance, the practice of business-based rotation is to let R&D staff rotate to other product-related business, including design, production, service, testing, and so on. In contrast, post-based rotation refers to the change in the positions of mid-level and high-level cadres. These rotations are mandatory every three years.

Huawei's job rotation system has several potential advantages. First, it reduces coordination problems among organizations. The job rotation system helps managers get familiar with other departments and facilitates cross-department cooperation. When the executive is eventually promoted to a more senior position, they would have become familiar with the different parts under their purview and found it easier to deal with the challenges at a higher level.

Second, it reduces the formation of interest groups in the organization. If managers stay in one post long enough, corruption will likely develop secretly, as in imperial times. Rotation reduces the likelihood of this happening.

PRESIDENT XI JINPING'S CV

The Chinese political system relies heavily on rotating promising officials before promoting them. Every leader since Deng Xiaoping has been promoted gradually after serving multiple roles at each level of government.

Before entering the highest echelon of the Chinese Communist Party, President Xi Jinping had governed the province-level jurisdictions of Fujian, Zhejiang, and Shanghai, with a total population of more than 120 million.

Xi started his political career in 1979. In 1982, he was sent to the County of Zhengding in Hebei province, where he was promoted to party secretary in 1983 and served that role for two years.

From 1985 to 2002, Xi Jinping spent 17 years in Fujian province, at that time one of the most economically vibrant provinces but also the most difficult to govern. There he started as deputy mayor of Xiamen, right across the strait from Taiwan. Xi then rotated through the municipalities of Ningde and Fuzhou, before being promoted to be the provincial governor in 2000. In Xi's 17 years in Fujian, the GDP of the province grew by 22 times.

In 2002, Xi was transferred to Zhejiang province, another economic powerhouse (and where Alibaba is based). Five years later, after accumulating enough experience, he was made the party secretary of Shanghai, China's largest and economically most important city, paving the way for his eventual promotion to the central leadership. ▪

Third, the job rotation system also provides some employees who are not progressing well in their current position an opportunity to try something else.

In addition, Huawei's job rotation system requires employees in its global markets to rotate across different countries. Many executives transferred from domestic to overseas units will face a novel, unfamiliar environment—a significant challenge but also at the same time a great learning opportunity.

We believe that the premise of the rotation system demonstrates that companies are very confident in their corporate culture and their products. Why?

Because the costs of job rotation are extremely high. We cannot expect too much of an employee by asking them to fit in a new environment quickly. If they don't fit in, the company has to pay for this wrong decision. As an old Chinese saying goes, one careless move forfeits the whole game (一着不慎满盘皆输).

For most people, forced rotation is certainly not a comfortable experience. However, in addition to the objectives mentioned previously, an additional benefit of the rotation is that it creates a common "memory"—a form of camaraderie—among the managers who have been through such experiences. We have seen many cases of managers complaining about the discomfort in the rotations, but eventually they become very proud of the organization they are working for.

Many internet companies, while not having a mandatory rotation system such as Huawei's, encourage their managers to rotate through different roles. Quite often, the biggest perk of undergoing rotations is that you get promoted faster.

Overall, through rotation systems, these firms build a pool of executives who have experience in decision rights across different levels and different units, access big-picture information, and understand how resources are allocated in the organization.

Internal Competition

Many Chinese internet firms deliberately create internal competition when resources allow. A key reason is to avoid making costly strategic mistakes, like the case of Alibaba's three entities (eTao, Taobao.com, and Tmall) competing to develop an ecommerce platform (discussed in Chapter 2) and Tencent's three teams working on WeChat at the same time.

Another important reason is to let the internal competition stimulate the organization, to keep its people alert, especially when the organization is seemingly successful. Instead of staying in the comfort zone, internal competition helps to push the organization to step out of the areas where they have had market dominance.

Tencent's internal competition was not limited to WeChat. In fact, different studios are competing for the company's billion-strong user base of gamers. Even globally in the MOBA (multiplayer online battle arena) space, Tencent's own *Arena of Valor* competes against Krafton's *PUBG Mobile* and Garena's *Free Fire*—both publishers count Tencent as a major shareholder. ByteDance does something similar.

Internally Developed Productivity and Collaboration Tools

Another feature that is probably unique to especially large Chinese internet firms is the adoption of internal productivity and collaboration tools. For example, Alibaba developed DingTalk, Tencent has WeChat for Work, while Byte-Dance has Lark.

PONY MA'S GRAY MECHANISM AND TENCENT'S INTERNAL COMPETITION

Pony Ma once famously proposed the concept of "gray mechanism"; that is, that a product should not be the way it was originally designed, but should stay in the gray area. In this gray area, users will decide the fate of the product.

Based on this philosophy, Pony proposed the following: tolerating failure, allowing moderate waste, and encouraging internal competition and internal trial and error.

WeChat's emergence from the competition among three separate teams, as discussed in the Leadership chapter, came through the gray mechanism. When the product(s) was being developed, the company did not provide a clear design idea or development direction. Rather, three teams worked independently and the users were the final judges of which product would prevail. ▪

In all these cases, the collaboration suite is not only used internally but marketed to external organizations, including smaller tech firms and more traditional organizations.

There is speculation that these companies build their own collaboration tools because they are competing against each other, making it not secure for anyone to adopt the tool developed by another. However, we believe a bigger reason is that because these organizations are constantly evolving and embracing change, it is difficult to find an external tool that caters to their specific, and constantly evolving, needs perfectly.

These collaboration tools have also evolved to reflect the culture, organizational characteristics, and workflow design of their parent groups. For example, DingTalk allows managers to ping and track their subordinates around the clock, often to the great annoyance of the employees. You would see Alibaba executives at casual lunches suddenly leave their tables because they are "dinged" by their boss.

In comparison, ByteDance's Lark focuses a lot on the needs of an "app factory," where collaboration and the use of OKRs (a goal-based management framework popularized by Google) are embedded into the system. Lark salespeople sell not only the tool, but ByteDance's way of effective collaboration.

Compared to Slack, Google Suite, or Microsoft Teams, the popular tools in the United States, DingTalk and Lark are much more comprehensive: they include not only messaging, documents, spreadsheets, and a calendar, but also video conferencing, task management, HR processes, automation, and much more. In other words, they try to cover every aspect of a modern worker's productivity and collaboration needs.

For Chinese executives, perhaps the most natural tool is WeChat for Work, for the simple reason that it resembles the consumer version of WeChat, which most of them already use on a daily basis.

Small, Agile Teams and the Attempts of Zhongtai

Chinese internet firms have their own evolution process. While the big decisions tend to be centralized at their headquarters, some large Chinese firms with multiple product lines have also tried to delegate some decision rights at the product/team levels.

This is especially true for some businesses, such as the gaming development of Tencent. Supercell, a gaming company acquired by Tencent, has developed several popular games, including *Clash of Clans* and *Clash Royale*. When asked about the secret of the company's success, the CEO of Greater China replied, "small teams and independence is the key to our game development."

Indeed, Supercell's teams have always been around 10 people—similar to Amazon's philosophy where a functional internal team needs to be so small that it can be fed by just two pizzas. Each team has the freedom to decide what kind of game they want to develop and ultimately whether they want to release it to market.

It should be noted that this is a perfect example of Tencent's game business. In fact, most of Tencent's game research and development teams adopt a small, project-based organization structure. The small team is a closed loop, which includes all the functions needed to develop a game such as planning, artwork, programming, and other functions. These kinds of small teams minimize the need for cross-departmental coordination on a day-to-day basis.

In addition, such teams can be reassigned or redistributed if the project they are working on becomes no longer relevant, or another more promising project needs more resources.

This arrangement avoids promising employees getting stuck in a part of the organization that is a dead end. However, if not executed well, it could also be problematic because some of the essential but not lucrative projects might find it hard to get enough staff.

TENCENT: "EVERYONE IS A PRODUCT MANAGER"

In the early days, Tencent's external partners and clients were often confused by their counterparts at Tencent because it seemed that everyone was carrying the title "Product Manager" on their business card.

A senior executive of a state-owned enterprise in Singapore, which had partnered with Tencent, complained once that they could not find the right protocol when structuring formal meetings. "When Tencent people meet our project executive, they send a product manager," he murmured. "When they meet our CEO, they also send a product manager."

However, that was an exact reflection of how Tencent's organization worked. Small, fully functional teams were organized along product lines—and of course the leader of such a team is a "product manager."

Tencent has recognized that confusion, and lately its external-facing executives have started carrying more conventional, easy-to-understand titles such as "vice president" or "director." ▪

To enable and empower such small teams to function well, especially in front-line product development, in recent years a number of large internet companies have tried to build a strong middle platform (Zhongtai).

Zhongtai, spearheaded by Alibaba but also tried by other companies, actually borrows the concept from the modern military. For example, during the World War II, the US military adopted divisions and brigades as its main combat mode. But in recent conflicts in the Middle East, the US military changed their tactics and used small teams of 7 to 11 soldiers at the front lines. Why? Because they are supported by a well-functioning, strong central system that provides air, missile, and intelligence cover whenever needed. As a result, commanders of small units are able to make very flexible decisions based on the battleground situation, which leads to optimal outcomes and minimal losses.

Similarly, Zhongtai is established to provide data, technology, and other resources to the front-line product teams. Zhongtai empowers and enables small teams on the front lines to make quick responses to the market and act nimbly.

Compared with traditional hierarchy, Zhongtai is in theory a better way to coordinate and support all the departments within the company. Therefore, the concept of "big middle platform" and "small front-line team" has been adopted by other Chinese internet firms, although they may use different terms.

For instance, Tencent reconstructed its business units in 2018, aiming to achieve a similar goal. The most prominent change in the organization was the establishment of a technical committee to provide support for other business units.

ByteDance's front-end platform consists of the company's products (apps) which are directly facing users such as Toutiao, Douyin, TikTok, Xigua, and so on. Each product needs to respond quickly to users' needs and pursue rapid innovation and iterations. Its back-end platform consists of the infrastructure and data center, where stability always come first.

Zhongtai (Middle Platform) is composed of ByteDance's technical modules or pieces that have been built for previous initiatives. Zhongtai provides services and support for all relevant products while facing technically repetitive tasks. For example, user behavior detected in Toutiao can help refine the recommendation engine, which also applies to Douyin (TikTok). While initiating a new project, Zhongtai can provide general technology, data, algorithms, and other solutions to speed up the development process as well as reduce costs.

However, it is worth noting that the success of Zhongtai is not guaranteed, and is still contested. In the military, although ground situations change rapidly, scenarios can be standardized and modeled, with the relevant systems developed. In the tech sector, however, there is much more complexity and much more unknown in scenario planning, which makes building a functional Zhongtai much more difficult. In addition, Zhongtai is often planned and released centrally, that is, by people who can't "hear the gunfire," so how to ensure that it can reflect the various product lines and markets is an acute challenge.

 ## CHALLENGES OF ORGANIZATION IN OVERSEAS VENTURING

Remember at the very beginning of this book, we mentioned Jack Ma's earlier saying that eBay and Amazon, as sharks, will find it difficult to compete with the local crocodiles once they are in the Yangtze River?

Many Western internet companies failed in China because their organizational structure was not able to adapt and align resources quickly enough to respond to changing market conditions as well as strong competition from domestic players.

When Chinese tech firms enter international markets, sometimes they face the same issue. More sophisticated and advanced organizations in China are not necessarily fast-adapting and nimble in other markets—even though they have the money, people, and resources act, they are slowed down by their organizational baggage.

Lazada versus Shopee

Let's use the example of Lazada versus Shopee, two companies we briefly discussed several times earlier in this book.

Lazada, a pan-Southeast Asia ecommerce platform acquired by Alibaba in 2016, was a centerpiece of Alibaba's ambition to become a global leader in ecommerce. To ensure Lazada's success, the group poured in a lot of resources: hundreds of engineers came in to rebuild the tech stack and product to be futureproof; Taobao, Alibaba's domestic Chinese ecommerce platform, would supply the sellers and product assortment, which was lacking in Southeast Asia; Ant Group would come in and help process payment and build fintech capabilities; Cainiao, Alibaba's logistics affiliate, would help coordinate and build the modern, efficient fulfillment infrastructure.

In comparison, when Tencent-backed Garena, a Singapore-based gaming company founded by three immigrants from China, decided to get into ecommerce, nobody from outside was convinced it could compete against the almighty Alibaba. A few friends approached by Garena in its early days to join Shopee, as the new ecommerce platform would be called, rejected the offer, which, five years later, would have made them multimillionaires.

Yes, five years down the road, Shopee, which started with much poorer products, product assortments, and fulfillment experience, is much ahead of Lazada in size, market share, and growth momentum.

How did this happen? The organization provided the biggest clue. When Lazada carried such big hopes for the group, and so many different departments and business units came in to support it, its organization became overly complicated.

For example, payment was handled by Ant Group, which might itself not be completely familiar with the market and payment needs: Who would have the final say over key decisions? And when implementation falls short of expectations, whose responsibility should it be? How do you align the KPIs of working-level people, since they report to different bosses?

Also, for decision making and resources planning, when executives in Singapore and Hangzhou are involved, how do they have the proper understanding and respond fast enough to the changing competitive landscape in Thailand and Vietnam? Do they fully trust the team in Thailand and Vietnam to make the most optimal decisions?

And when Zhongtai was introduced to Lazada, how would the tech decision makers in China make sure all the potential product issues and scenarios in Southeast Asia are considered?

In comparison, what Shopee does is not magic. It has a vision and a plan, and it executed this plan relentlessly. More importantly, it has a functioning central command in Singapore, where many of the key decisions and resource allocations were handled in a top-down manner. Its leaders spend a lot of time personally in key markets such as Indonesia (and now Brazil, with its Latin America expansion), and are able to make decisions and adjust resources quickly.

As a result, Lazada, with its superior people and resources on almost all fronts, gradually ceded its market leadership position to Shopee. Now, when you have a competitor ahead of you, it is more difficult to make the right decisions—an ex-executive told me that Lazada's strategy has been flipflopping multiple times a year, between "we need to do the right thing for the long term" and "we need to catch up with Shopee's GMV [gross merchandise value], fast."

Shopee is an exceptional case as its founders are Chinese migrants and are more aware of the ecommerce model, as well as Alibaba's strengths and weaknesses. To be fair, if Shopee was not around in Southeast Asia Lazada would probably have been able to grow its market at a more comfortable pace, and solve problems along the way. However, when you enter big markets, the last thing you want to assume is that there will be no strong competition.

Adaptation of Organizational Learnings Overseas

The Lazada versus Shopee competition is a good reminder to Chinese tech leaders. The organizational elements that evolved over time and work for the Chinese domestic market are the results of natural selection—the survival of the best fit to the environment. However, when you copy and paste this existing organizational structure to the overseas market, you are very likely to encounter challenges, either because of logical inconsistency in a specific market, or failure in implementation due to lack of key ingredients—people who had grown in the same cultural and organizational context.

Overall, venturing overseas is more difficult and complicated than entering a new province in the China market or building a new business line. How to succeed in a market where culture, context, and user background are completely unfamiliar is an important issue for Chinese firms' overseas teams to consider. Any product strategy needs to be supplemented by its organizational structure. The fundamental question of task division and integration merges again, and becomes more difficult to address because of different cultural context.

Let's get back to the three fundamental issues in organization discussed in the first section.

Decision Rights

It is not difficult for Chinese internet firms to understand that an overseas business needs an overseas team to operate such that they can make judgments and decisions timely. The concept of empowering the local teams, or the idea of decentralizing decision-making rights, has a precedent in Chinese history.

During China's Civil War, Mao Zedong told his commander Su Yu three times that he did not need to ask the headquarters for instructions and could take direct actions. Later, Su commanded his troops and won a series of victories in a series of classic battles, which made the Communist Party army defeat the Kuomintang army. Needless to say, the success of the Communist Party was inseparable from Mao's decentralization of decision-making power.

The Art of War, an ancient Chinese military book written by Sun Tzu, states that if the general is at the battlefield, he does not have to obey all the emperor's orders ("将在外，军命有所不从"), which means that frontline soldiers do not have to report to headquarters and can make their own decisions.

After all, the breadth and depth of information about the local market that headquarters can access would not be comparable with that of the local team. If local teams don't have discretion in decision making, they cannot adapt to and respond to the local market quickly—a recipe for failure.

However, the devil is always in the details. For instance, in the decentralization structure, what are the business and operational scope of the local teams? Of what kinds of decisions can the local team have the "real" decision rights and shoulder the responsibility?

A key question here is whether the overseas teams should have control over tech and product. We discuss a mini case—WeChat in India—in Chapter 7, Product, which clearly reveals its product design problems in the local market. Even if the local team understood it accurately, they had to communicate with

LALAMOVE AND ITS POKER PLAYER FOUNDER

Hong Kong-based Lalamove is an on-demand marketplace for cargo owners to book small vans. It addressed a key issue of small businessmen and distributors: How do I get small quantities of goods delivered quickly?

Its founder, Chow Shing-yuk, is legendary. A graduate of Stanford, he made his first pot of gold as a professional poker player in Macau for almost eight years. After spotting the Uber-like opportunity in Hong Kong, he started a company called EasyVan, which was later renamed Lalamove.

Perhaps because of his poker instinct, Shing resisted the temptation to burn money to acquire customers—which many of his competitors did and died subsequently. Instead, Lalamove focused on a simple strategy: making their stickers visible on as many vans as possible. This cheap marketing strategy, executed consistently and relentlessly, gave Lalamove a really good edge in customers' mind as the go-to place for on-demand logistics.

The early separation of Lalamove's international app from its Chinese version, Huolala, also contributed to the former's successful expansion across Asia and Latin America. ▪

tech and product teams in China, often not a very effective process. Why would the product team prioritize your requests over others, especially those coming from a team in the much larger domestic market in China?

In this regard, TikTok and the Hong Kong–headquartered on-demand logistics platform Lalamove did a much better job. Both separated their international from their Chinese products very early on, with separate apps, database centers, and teams working on different lines of businesses. That already makes it much easier for local teams to respond more nimbly to the international markets.

Information Communication

Our interviews found that there are general problems in communication from the overseas teams to the firm's headquarters. One reason is, ironically, because of the Chinese firm's advantage in "embracing change" discussed earlier.

Compared to Western peers, Chinese firms, as we discussed earlier, tend to have continuous restructuring. Needless to say, each organizational reshuffle means a drastic shake-up, making it difficult for the overseas team to build regular contacts with the head office in order to communicate the information.

Indeed, people we interviewed pointed out that their headquarters' organizational structure is constantly changing, so they cannot find the corresponding person in the headquarters to report the situation in a timely manner, let alone ask them for resources.

In addition, the communication problem is not only in the bottom-up process, but also occurs from the top down: when headquarters switches the strategic direction, the same information may not communicate to managers in the overseas markets in a timely manner. That can also create big problems.

An example was Baidu's O2O (food delivery and other local services) plan in Brazil. When a strategy was set and communicated to the local team, they started working on it, signing with partners, sending messages, and planning for the launch.

However, due to the changing environment in China, the headquarters decided to shift the priorities. With no prior notice, all the efforts by the overseas teams backfired: all the promises to local partners could not be delivered on.

It was a lose-lose situation. The local team was stuck in the middle, facing an awkward relationship with local business partners; the firm lost its reputation in the market.

We have also seen the situation when Ant Group's IPO was shelved, and country managers of the group in various foreign countries were summoned by regulators to figure out what was really going on and whether their commitment to the specific markets still stood. Most of the country managers could not answer because whatever happened in China happened suddenly, and in the middle of the pandemic it was difficult for them to work out a solution.

Even under normal circumstances, communications could be a hassle. As we mentioned earlier, especially in emerging markets information is often not complete or available. Even if you have sourced very detailed intelligence it is often difficult to decide what to trust or use. Decisions having to be made at headquarters or jointly often leads to suboptimal outcomes.

Resource Allocation

Without matching resources to the overseas units, decisions are just a piece of paper. Drawing the organization chart showing the resource flow from one unit to another is simple, but the reality is far more complicated.

While internal competition for resources exists in any organization, overseas units of Chinese internet firms faced two additional challenges. First, the domestic market in China is too big, which also means that the overseas market in the initial stage is perhaps too trivial—even with the leader's commitment, sometimes managers running overseas have to fight much harder to get the resources they need.

One executive we interviewed said that, despite sales growth of 200% in the Thailand market, it contributes less than 1% of the group's total sales. It simply cannot get enough managerial attention from the various departments that are supposed to contribute resources.

Second, *guanxi* (connections and relationships) is deeply rooted in Confucian doctrine. In Chinese companies, guanxi plays an important role in getting resources and, more important, information about what is really going on. If interpersonal relationships are not handled well or your guanxi with other senior colleagues is not close enough, there may be obstacles when you try to secure the support and resources you need. You might not even know how decisions are made and who influences decisions at headquarters.

Shopee did not have that problem. In contrast, resource allocation from Alibaba's headquarters slowed the functioning of Lazada drastically. How Shopee prevents the same problem from emerging when it transitions into a global company with operations in Europe and Latin America will be very interesting to watch.

BREAKING THE SILOS AND HURDLES

Fundamentally, the organization challenges we discussed earlier could be pinned down to how to integrate at higher levels when some decision rights are pushed down to the local managers across different departments.

Some companies have tried different approaches to tackle the organizational challenge in overseas markets: Tencent has an international business group (IBG) where the headquarters and overseas teams have a channel to communicate. Huawei used a dual-leader system in foreign markets: one is a Chinese expatriate from the headquarters who understands Huawei's culture, product, and organization, and the other is from the local market and understands local culture, market, and connection.

These are often not ideal or adequate, but they are part of the learning process. Organizations will need to be constantly evolving to adapt to the realities of a multinational business. Two critical enablers of improvements here are leadership and people, which we discussed in the previous two chapters.

Eventually, large Chinese tech companies will become mature and more MNC-like, and growth will not be as fast as they have been enjoying. In navigating their organizations toward maturity, leaders need to be acutely aware of the organizational challenges that come along the way.

One saying that is popular in China now probably summarizes the considerations needed for organizations: "If the dragon slayer lives long enough, he will eventually become the dragon."

 ## QUESTIONS TO CONSIDER

To conclude this chapter, we propose the following questions for you to consider if you are exploring the overseas market.

- How would you make key decisions for the market? Who has the right to give input, and who has the final say?
- Would you have the right information and communication flow in the organization when the expansion is happening? If not, what are the bottlenecks, and do you have the right means to fix the problem?
- How do you intend to allocate resources to the expansion market, especially shared resources such as supply chain, product, and tech?
- When your organization undergoes structural changes in your home market, how do you intend to maximize the positive and minimize the negative impact on your expansion markets?
- Operating in different markets, how do you ensure that the organization is agile and competitive?
- How do you ensure that you have a backup solution immediately if the organization for expansion fails?
- What are all the organizational roadblocks for a successful expansion? How do you remove them?

NOTE

1. Phanish Puranam, *The Microstructure of Organizations* (Oxford University Press, 2018).

7

Product

Everyone in a tech company needs to think like a product manager.

—Anonymous Chinese internet participant

S O FAR WE HAVE discussed leadership, people, and organization, all of which are core capabilities and competences a tech company in China needs to have in order to grow while withstanding tough competition.

Eventually, all these will need to be translated into products that customers use and pay for. Product is the last piece of our POP-Leadership jigsaw, and interestingly, the part that Chinese internet companies are most confident about in their journey overseas, for some good reasons.

Many Chinese internet products have withstood the test: as of January 2021, WeChat, Tencent's popular social app, has 1.09 billion daily active users; 630 million users have transacted on Meituan's food delivery and other platforms over the past year; and more than 600 million people use TikTok's Chinese version Douyin on a daily basis; Alipay supports more than 1 billion users and 80 million merchants.

Not only the demand but also the fickleness of Chinese consumers have forced these products to evolve constantly. Even some of the smaller tech products, through competition in their own niche sectors, have evolved to be globally competitive.

The product we describe here is not only the tangible offerings, but also intangible services and business models. It is the central vehicle for the companies to interact with the users, deliver value to them, and meanwhile capture value from serving the customers. In addition, a series of decisions will need to be made around the product, such as what, when, where, and how.

For instance, if you were heading a Chinese internet company's overseas journey, would you use the existing product and business models that have proven to work well in China, or would you emphasize local adaptation to cater to different markets?

Where would you start your expansion, in developed countries or emerging markets? If you are going to target emerging markets, would you bet on your chances in Southeast Asia, the Middle East, Africa, or Latin America? If in Southeast Asia, would you choose Indonesia, Singapore, or Vietnam, or perhaps a few of them simultaneously?

Relatedly, when is the best time to enter a market? Would you do it sequentially or enter all markets simultaneously? How would you enter? Given the resources you had, would you partner with local players, acquire existing local players, or build the business from ground up in destination markets? And how would you achieve your goal?

The question about overseas expansion is, How can Chinese companies adapt all their experiences and sophistication to a new market? In this chapter we examine some of the key issues.

Let's start by addressing a question many are curious about: Why is every Chinese consumer app a multivertical super app?

WHY IS EVERY CHINESE APP A SUPER APP?

When emerging major consumer tech companies in developing countries, such as Grab in Southeast Asia, Gojek in Indonesia, Paytm in India, started branding themselves as "super apps," they all pointed to China.

As they explain, consumer tech companies in China first adopted and perfected the super app methodology: WeChat is not only instant messaging, but also gaming, content, payment, and financial services; Ant Group's Alipay has all kinds of financial services on top of its payment offering; Meituan allows customers to not only order food delivery, but also use power banks, share bicycles, and many other things.

Even on the productivity suite front, those developed by major Chinese tech companies, such as DingTalk, WeChat for Work, and Lark of ByteDance, discussed in the previous chapter, are much more comprehensive in features and functions compared to their Western counterparts.

However, if you look a little closer, you will realize that none of these consumer tech giants in China uses the term "super app" or anything similar to describe themselves. For them, it seems natural to have evolved to where they are today: providing multiservice apps to consumers, with a lot integrated in the back-end.

Why is that the case? Why wouldn't the Chinese equivalent of Facebook (WeChat) stick with social media and content? Why wouldn't the Chinese equivalent of Amazon (Alibaba) focus on ecommerce? Why would everyone want to encroach on each other's space?

There are, in fact, two reasons for this, and they are interrelated.

First, as a Chinese tech leader once pointed out to us, "Every successful consumer tech company in China is at its very core a master of consumer traffic." This means that companies initially achieve success by being able to acquire and retain customers more cheaply than their competitors. The ultimate formula is that customer acquisition cost (CAC) must be lower than customer lifetime value (LTV).

In order to sustain the business, many major tech companies resorted to offering additional services to better retain the consumers, and also monetize more from each customer—and both aims eventually increased the customer LTV.

You do not simply lump everything together. Each new offering that is added to the app has to be closely related to existing offerings such that consumers find it natural to use the new service, and the app increases its conversion rate.

The second factor is unique to China. In the early days of tech development, Baidu, Alibaba, and Tencent banned each other's links in order to retain customers: for example, you can't share a Taobao product on WeChat, and you can't use WeChat as a payment method on Taobao.

The tactic, initially used to avoid leaking too many users to other services, effectively created a few walled gardens of ecosystems. WeChat started systematically channeling their customers to Pinduoduo and JD.com, two ecommerce companies Tencent had invested in, while Alipay made repeated attempts to introduce social elements in its app. Both Tencent and Alibaba have invested in companies across all consumer tech sectors to rival each other.

At the same time, smaller but emerging tech companies, fearful of becoming too dependent on the giants, also started to build their (smaller) walled gardens.

Therefore, de facto super apps proliferated. At the end of 2021, the government started making efforts to break these walled gardens and force interconnectivity. Whether that will make the ecosystem more open and companies more focused on developing their niche specialties instead of going multivertical remains to be seen.

START WITH IMPERFECTION, BUT IMPROVE FAST

All these Chinese multi-vertical super apps always started with an initial core feature that defined the product before expanding into other features.

And yes, from the customers' point of view, Alipay is the product. Payment, credit, lending, wealth management, and so on are all features of that product. This mentality is quite different from how large traditional companies plan things—in their mind, each of these features is a product.

By defining everything as a feature, tech companies inadvertently kept silos from being formed, and therefore customer data to be centrally managed and monetized.

Internationally, tech startups are advised to develop minimally viable product (MVP) first to test the market, before adding features. Major Chinese tech companies follow the same model. Pony Ma of Tencent once said, "The market will never be patiently waiting. In competition, a good product always starts from imperfection."

However, he added that once a product is in the market, it needs to make very frequent improvements toward perfection. "If we fix one or two problems every day, within a year we will have a very good product for our customers."

Jack Ma also advocated for "dream big, start small, learn fast," while other tech leaders have also emphasized the same logic of starting small and iterating fast.

Therefore, new products are often launched with the intention of taking advantage of opportunities to make improvements in the future. Those that withstand the test of the market will be improved and eventually become successful, while others might be phased out very quickly.

The deep strategic thought of leaders we discussed in Chapter 4, Leadership, will guide the company in the big picture as well as in direction; however, it rarely becomes a hindrance to fast product iteration. Jack Ma specifically said to avoid falling into this situation: "At night you think about thousands of different paths, the next morning you take the same old path."

Timing is also important in product considerations. While major tech companies are fighting in very fierce battles to not cede ground to others, many have advocated that being the first mover does not necessarily translate into advantages.

As Pony Ma said: "Never think that you can have peace of mind because you are first in the market. I believe in the age of internet, nobody is dumb. Your competitor will wake up fast and catch up even faster. They will even do it better than you. As a result, your moat can be broken into at any time."

LAUNCHING A PRODUCT OVERSEAS: THE QUESTIONS OF WHERE AND WHEN

The guiding principles discussed in the previous section are also adopted when Chinese companies expand overseas. The first decisions they need to make are where to go, and when.

Where

Chinese tech achieved such success in a very short period of time partially thanks to China's massive population, who speak the same language, have good consumption power, and are now connected to the mobile internet.

When companies go overseas, an instinct is to look for markets with similar traits. Therefore, India and Indonesia, two Asian markets with close proximity to China and vast populations (1.4 billion and 270 million, respectively), became key targets.

Baidu chose Japan as its first overseas market, as mentioned earlier; it later expanded to Brazil and a few other large markets. Didi, the Chinese ride-hailing giant, also chose Brazil and Mexico as the first stops for their global expansion. More frontier countries such as Nigeria, Pakistan, and Bangladesh became targets.

"We need to go where the consumers are," one executive shared. The population density map (see Figure 7.1), therefore, becomes a key reference point.

Using population density as the compass, we can understand why Ant Group is a big fan of Asia. The Alibaba-affiliated fintech giant has built an

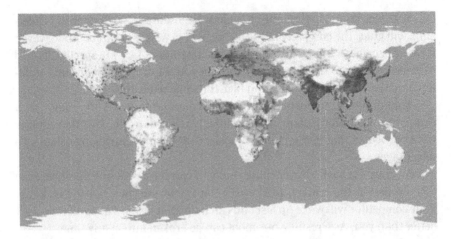

FIGURE 7.1 Population density map (Imagery produced by the NASA Earth Observations team based on data provided by the Socioeconomic Data and Applications Center (SEDAC), Columbia University.)

e-wallet designed for local markets and set foot in the Asia-Pacific region, including India (Paytm), Thailand (TrueMoney), Philippines (Gcash), South Korea (Kakaopay), Indonesia (DANA), Hongkong (AlipayHK), Malaysia (Touch 'n Go), Pakistan (EasyPaisa), and Bangladesh (bKash).

Southeast Asia, in particular, has become a hotbed for venturing Chinese tech firms, despite being a complex region with different languages, customs, and geographies. The reasons are simple:

- Big market: A combined population of 650 million, a low median age, a fast-improving infrastructure, and a strong willingness to consume.
- Geographical proximity: In the same or similar time zones, with short air and freight distances.
- Historic and cultural proximity: Trade between China and Southeast Asia dates back millennia, and most countries boast a large, ethnic-Chinese business class.
- The existence of Singapore: An ethnic-Chinese majority business hub that many Chinese companies have used to set up their global headquarters.

We also notice that Chinese companies, big or small, tend to form clusters when they are outside China. This practice dates back to imperial times, where migrants from the same hometowns tended to move to places where there was already a community of fellow countrymen: Cantonese in

California and Peru and Malaysian capital Kuala Lumpur; Fujian people in Singapore, Indonesia, and the Philippines; Chaozhou (Teochew) natives in Vietnam and Thailand; and Hakka in India, Mauritius, and Borneo.

Companies operating in foreign markets are not always able to tap into the most trustworthy local sources. A congregation of the existing Chinese community often facilitates the flow of information and fosters infrastructure. For example, the clustering in Brazil of gaming companies was because a set of service providers figured out payment, customer acquisition, and operations for Garena's Freefire, while service providers camped in the same housing estate (Taman Anggrek) in Jakarta, Indonesia, provide a whole suite of services for fintech companies, from legal and accounting to credit assessment and debt collection.

However, the clustering effect also increases the intensity of competition among Chinese companies. JollyChic and the ecommerce industry in the Middle East is a good example. JollyChic was the first Chinese ecommerce company to "discover" the market of Saudi Arabia and invest heavily in that market. In 2016 and 2017, it enjoyed almost unrivaled success in offering large selections of fashion products into the Kingdom. However, once its success became known through the Chinese communities in Dubai, many other Chinese cross-border ecommerce players jumped on the bandwagon. Very quickly, the blue ocean of cross-border ecommerce for the Middle East turned into purple or even red.

When

Every Chinese leader recognizes the importance of timing, because many successful ones benefited from the explosion of mobile internet users in the early 2010s. They are also acutely aware that, as we quoted Pony Ma earlier, being the first in the market does not mean you can have peace of mind.

First-mover advantage does exist in some situations. When you occupy the majority of customers' mindshare and brand recognition, it will take a challenger much more effort to dislodge you.

Baidu encountered that in its expansion into Japan: while its Japanese search product was honestly quite good, Google was already established there. It was very hard to beat an incumbent using exactly the same method.

Similarly, when WeChat lost its initial battle against Line—a Japanese/Korean chat app—it was nearly impossible to mount another challenge. All the users have already formed their social relationships on Line, which becomes very difficult to disrupt unless Line makes a big mistake, or a new channel opens up (as smartphones did in the 2010s).

But, *first-mover advantage only works when first movers are able to capture the market and build a strong moat around them.* The best moat, in our opinion, is a stabilizing market where upstarts do not find the opportunities to launch an assault and topple the incumbent. Otherwise, being the first mover does not warrant you the advantage.

The story of Lazada versus Shopee, which we described in more detail in Chapter 6, Organization, is a good illustration of this as well.

Lazada had all the elements of a classic first-mover advantage. It was the first major ecommerce company that managed to establish a leading presence in all six Southeast Asian countries. Its integrated logistics network had no rivals in scale and sophistication. It had the brand name and recognition as the Southeast Asian ecommerce story.

In fact, in 2016 when Alibaba was eyeing Southeast Asia, Lazada was the only rational choice to acquire. Now coupled with strong product, tech, and operational capabilities as well as deep ecommerce expertise from Alibaba, the competitive landscape in Southeast Asia's ecommerce scene seemed set.

However, in an emerging market or an emerging sector, there is a huge risk in being the first mover. The market is immature and uncertain, and your products do not necessarily fit the market because of that uncertainty. Even worse is the lack of infrastructure and customer education.

The first mover, in this instance, had to dedicate a lot of time and investment on the groundwork—creating the infrastructure and educating customers. In the case of Lazada, it had to build Lazada Express because nobody else at the time was capable of delivering ecommerce parcels efficiently and effectively; it also spent a lot of time and effort teaching merchants how to use ecommerce.

Therefore, seemingly having 70% of the market share is not a defendable position because the existing market is still very small, and the biggest opportunity is in future growth.

Shopee, in this case, swept in, offering merchants, who then already had a grasp of ecommerce, a second choice; and because it was a latecomer, it could be more flexible in product and operations, while Lazada struggled to make its existing product more agile. Lazada's heavy infrastructure and deep, entrenched collaboration with other Alibaba units (on payment, logistics, and cross border) also slowed it down, while Shopee did not have these "legacies."

Shopee made it, and it was not alone: Facebook was not the first social media platform (MySpace was more promising before 2009); Google was not the first search engine (remember Yahoo! and AltaVista?); Amazon was definitely not the first ecommerce platform.

In our opinion, any eventual winner would leave the market education to others, and go for the kill at the right time. As we quoted Duan Yongping of BKK earlier: "Let you start first, and I will go for the kill with the model you have explored, in a market you have educated. In your eyes, I am your competitor; In my eyes, you are my tool."

In fact, the first-mover advantage is an oversimplification of the complex issue of navigating the complex landscape of a nascent market. In our view, you only want to become the first mover when you are committed to pushing through the uncertain period of market education and are still able to scale up afterward.

Otherwise, it makes sense to learn from Shopee: get the resources ready and launch your product in a massive push once the leading player becomes exhausted but the market is yet to be fully occupied.

Shopee has pulled off this trick multiple times, and not only in ecommerce. Its payment affiliate ShopeePay only started ramping up in the second half

IS SHOPEE A CHINESE COMPANY?

A question often asked in Southeast Asia is: Is Shopee, or the Sea Group as a whole, a Chinese company or a Singaporean/Southeast Asian company?

Its founders, Forrest Li, Gang Ye, and David Chen, were immigrants from China who obtained Singaporean citizenship. The company filled its top ranks with experienced hires poached from large Chinese tech companies. Tencent is its largest shareholder.

On the other hand, Sea Group started in Southeast Asia, and has always been very close to the markets. Its strategic choices and execution, while exhibiting strong characteristics of a Chinese tech company, also demonstrate deep regional understanding and capabilities.

Unlike Lazada senior managers sent from Alibaba who spend most of their time in comfortable Singapore, Sea's leaders are constantly on the ground in Indonesia, Thailand, and Vietnam. Their connections to China allow Sea leadership to see what is possible (and proven); at the same time, their regional roots allow them to execute much better than their Chinese peers in Southeast Asia.

As a friend who joined Sea from a major Chinese tech firm said, "We feel that the pace here is very slow; however, our regional colleagues are telling us it is already much faster compared to other regional tech firms. They feel they are working in China." ▪

of 2019, after competitors OVO, Dana, and GoPay were exhausted after very expensive subsidies to educate the market; its recent foray in Latin American markets (Brazil and Mexico) also followed years of buildup by AliExpress, Alibaba's cross-border affiliate.

If being the first mover does not guarantee you the advantage, the question becomes "When is the right time?"

Based on various discussions with different executives, we believe that there is an optimal window of opportunity when the market (infrastructure and customers) is ready but has yet to take off (Figure 7.2).

This window can open *when there are dramatic shifts in demographics in the market.* Such shifts include new generations of consumers or new forms of access—an obvious example is when millions suddenly gain access to the internet through smartphones.

It can also open *when competitors make a strategic shift.* For example, Pinduoduo seized the opportunity—created when Alibaba moved upscale (for better profitability)—to attract the leftover market for cheap products and low-end customers. When it built volume in these markets, it also gained a cost structure designed for low-margin markets and an ability to scale up, capabilities that incumbents find hard to match. It then moved up the value chain to attack the more profitable segments guarded by incumbents. Sun Tzu summarized this well in *The Art of War*: 善用兵者,避其锐气,击其惰归 (The strategist should avoid the enemy when he is strong, and attack when he is tired).

To be able to detect such opportunities, companies and leaders need to have a very acute sense of the movement in the market; to seize these opportunities, they need guts and commitment, as we discussed in Chapter 4, Leadership.

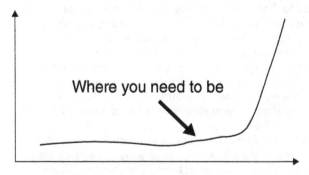

FIGURE 7.2 Where do you need to be?

In addition, they also need to be ready in both product and organizational capabilities, and have a support ecosystem in the market to tap into. Otherwise they can see the opportunity slip away in front of their eyes.

The ancient Chinese philosopher Menicius summarized it all: 天时地利人和 —which literally means, right time, right place, and right people. To seize the right window of time, the key is not only patience, but also a deep understanding of the market, as well as yourself—there is no shortcut for that.

◼ LOCALIZATION: THE QUESTION OF WHAT

Localization of products is often the most talked about subject in any forum on Chinese tech companies' overseas ventures. While we believe the question of "localization" should really be a question of "understanding" and "adaptation" of leadership, people, and organization, eventually the product offered in specific markets needs to fit with what the market needs and is willing to pay for.

A simple question here is about the balance between standardization and localization: To what extent should the products be customized to local market needs?

The tradeoff is easy to understand. Global standardization, using the same product (brand, design, features, and operations) across multiple markets, helps reduce cost, create synergies in centralized R&D, production, and operations, whereas localization—the process of adapting product design, brand, format, operations, and marketing to specific characteristics of the destination markets—creates better-suited products and thus enables business to grow better locally.

Earlier academic and nonacademic research has studied this topic extensively, with case studies covering prominent companies such as Walmart, KFC, and P&G. The recommendations of such studies are usually a mix of both standardization and localization.

The challenge for tech/internet companies is that they are not able to afford the years that traditional companies took to understand a market and calibrate products. Because the distribution to end consumers is now done via smartphones directly, the timeline is much condensed, market feedback is almost instantaneous, and market share can be gained or lost quickly.

Internet companies, as a result, need to adapt their products much more quickly compared to traditional companies. This is in their nature and one of the key reasons they stand out in their home markets.

However, in foreign markets, this adaptation has proven to be quite difficult, especially for companies whose home presence is big and successful.

We have to say that US tech products are usually less localized, especially to emerging markets where infrastructure, payment methods, and customer trust are all different. That is probably why Amazon still concentrates its operations largely in developed markets (with the notable exception of India) and Facebook focuses on advertising, without getting into ecommerce, which involves the supply chain.

Lack of localization is often identified as a key reason why eBay, Amazon, Yahoo!, Groupon, and others failed in China. However, the reality is that, because China is a vast market, it naturally attracts capital and entrepreneurs, and you will naturally have a lot of well-funded local competitors who understand the market better and can move much faster compared to global firms. We all know that if you iterate once every month while your competitor iterates once every week, you will fall far behind within a year's time.

Chinese tech companies, in their expansion overseas, especially in large markets, face the same reality. Ant Group faced challenges in adapting its advanced financial technologies with its joint ventures in Southeast Asia, losing to much less sophisticated local rivals; Baidu struggled to make its product click with Japan's and Brazil's consumers after multiple attempts and acquisitions.

In the case of Ant Group, it is not surprising that they are proud of their technology, the mobile payment system, the cloud infrastructure, and their capability to integrate with various pieces of Alibaba's ecosystems.

However, in the perspective of consumers and business partners in Southeast Asia, the number of total sales in the "Double 11" online shopping festival in China is something they will never reach, and thus matter little to them.[1] Ant Group's technology and offerings, while advanced, are too complicated for their needs.

Similar to US companies in China, many Chinese tech firms struggle with reconciling their large and successful products in China with feedback from other markets that sound counterintuitive to what they are used to.

We have seen engineers deployed to Ant's Southeast Asian ventures struggle to understand why people do not use QR scan to pay, very similar to early PayPal executives who did not appreciate why people did not use credit cards to pay.

Ironically, this is the "trap of (prior) success," or, as people in China would call it, pathway dependency. It is a systemic people and organizational mind shift that is needed, not something that product can solve alone.

Interestingly, small or less successful companies in China often do not have such baggage and therefore are able to move much faster to localize their products in target markets. Of course, the fact that their leaders can spend substantial time in those markets also helps drastically.

Transsion, a smartphone brand not known in China, has had great success in the African market. The Shenzhen-based company tailored its phones to provide functions in response to customer pain points that are unique in Africa. These efforts help to distinguish its smartphone from other brands and provide exceptional value to customers.

HOW TRANSSION LOCALIZED IN AFRICA

- Transsion launched the first dual SIM card phone in 2007, and a four SIM card phone in 2008. Given the fierce competition between telecom operators in Africa (e.g., Telkom, Vodacom, Orange), call charges between different operators were expensive. To take advantage of cheap intra-network rates and promotions, users needed more than two SIM cards; otherwise, they had to change SIM cards frequently.

- Acknowledging the importance of music and entertainment, Transsion developed louder and clearer speakers for music and phone calls.

- To overcome the problem of insufficient, unreliable electric power supply in some countries, Transsion developed a long-lasting battery and low-cost quick-charge technology.

- To take account of language differences, Transsion developed operating interfaces with local languages, such as Amharic Tigrinya and Oromo in the horn of Africa, Swahili in Kenya and Tanzania, and Hausa in Nigeria.

- To cope with heat and humidity, Transsion developed an anti-corrosion coating technology tailored to hot climates, with strong acid resistance up to pH 3.5, with a sweatproof algorithm to improve the success rate of fingerprint unlocking, as well as a sweatproof USB interface and heat protection for electronics.

- Its most well-known function was the "beauty camera," which made it easier to take selfies for dark-skinned sub-Saharan Africans in low light—a feature lacking in mainstream brands. Transsion built a database based on millions of pictures of Africans and optimized the camera and algorithms to light up the faces adequately in photos taken in low light settings. ▪

However, when such Chinese companies successful in one market or region decide to expand, they face the same issues. Transsion explored Southeast Asia markets, but could not manage to penetrate them. JollyChic, the aforementioned Chinese commerce platform successful in Saudi Arabia, found it very difficult to enter Indonesia—their entire product and operations were so attuned to the Saudi market that executives found it very difficult to adjust—Indonesia customers (who are also Muslim-majority) respond to the same product so differently.

Many large Chinese tech companies also recognize the localization challenges and are taking active steps to address them. A key strategy is what ByteDance and Lalamove did, separating international products from domestic Chinese ones, allowing the international version (such as TikTok) to develop independently without the baggage. This is not only to respond to geopolitical concerns, which led to former US president Trump's threat to ban TikTok, but are also practical: a Douyin update will not affect TikTok and vice versa; the TikTok team can focus on building international content without interference from Douyin's massive trove of local Chinese content.

ByteDance even set up the TikTok headquarters in Singapore to further strengthen its product independence. The company is also trying to replicate this separation in other offerings. For instance, Lark keeps its productivity suite (which we mentioned in Chapter 6) separate from its domestic Chinese version, Feishu (飞书).

"China's market accounts for more than 90% of our business. Had we bundled them together," one former senior executive told us, "we would not be able to develop international markets properly because Chinese product requests would always take priority—and there are a lot of such requests."

Ant Group further defined the approach, by setting up a G-local (global localization) in its subsidiaries in Southeast Asia to systematically harmonize Chinese product experiences with local realities. And as we have always argued, such efforts need to go beyond product to organization and leadership.

WeChat's Failed Global Quest

WeChat, first launched in 2011 by Tencent, is a stunning success in China. On January 21, 2021, WeChat's 10-year anniversary, Tencent revealed that the number of daily active users (DAU) had exceeded 1.09 billion.

When WeChat was launched, Tencent already had a massively popular chat and entertainment application called QQ, which had 721 million active accounts (users can have multiple accounts) in 2011. QQ offers not only instant messaging, but social networking, games, and many other functions to attract and retain users.

In the mobile arena, WeChat had a similar, and much more successful, trajectory. Functions such as voice messages, video and voice calls, a feed for sharing photos and links, games, payment, content publishing and consumption, short-form videos, mini programs (de facto apps based on WeChat), and ecommerce and food delivery are progressively added. Through an open platform with numerous third-party partners, WeChat has created an ecosystem where users can fulfill most of their needs—an "app for everything."

Throughout the years, many friends, Chinese or foreign, have asked us, "Why isn't there something like WeChat in other markets?" or, more directly, "Why hasn't WeChat expanded successfully outside China?"

Well, it's not that WeChat hasn't tried.

Right after hitting 100 million users in just 433 days, WeChat decided to go global. India, whose population size was similar to China's, and where Tencent already had acquired a travel services portal called Ibibo, became a key market in which to test the water.

In May 2013, a major marketing campaign was launched, featuring several well-known Bollywood stars, including Parineeti Chopra and Varun Dhawan. The immediate results were promising—around 25 million new subscribers within a short period. WeChat was the top-ranked app in the Google Play Store in India for six straight weeks.

However, that success was short-lived. While a massive number of users joined, they did not stay. Retention became a serious problem that was never fixed.

In February 2014, Rahul Razdan, head of WeChat India (who had joined with the Ibibo acquisition), quit. One year later, WeChat wound up the majority of the Indian team.

On the other hand, WhatsApp prevailed in India, reaching more than 400 million active users by 2020.

India was not the only market where WeChat made a splash that did not stick. The same year, with the endorsement of famous Argentinian footballer Lionel Messi, WeChat scored more than 100 million users globally. Most of those users did not stick, either.

WeChat failed to replicate its success outside China. But why? They made the same mistakes that many US tech companies made when they entered the Chinese market. A key was the failure to localize their products.

Many analyses argued that the design principle of WeChat was to build up an "all-in-one social app" that provided various functions to its customers.

What users in India (and many other countries) really wanted was a tool to communicate with others quickly and simply, exactly what WhatsApp offered.

Compared to WhatsApp, WeChat was bulky, in both memory size and power consumption. The original WeChat app, with a size of 40 megabytes for the Android version, was too big for most smartphones in India, which had less than 200 megabytes of memory.

The patchy mobile internet connection at the time in India, as well as in many other countries, meant that WeChat would often crash (very frustrating for instant messaging users), while the WhatsApp user experience was much smoother. The poor (and pricey) network also meant that users could not send or receive large pictures/videos/files, or video chat, in India. Later on, (Chinese) file-sharing apps such as ShareIt and Xender, which used Bluetooth or a WiFi hotspot, became popular.

However, the size and "all-in-one" design were not the only factors. When it comes to building a social network, each culture is different. While WhatsApp chose the universal principle of simplicity, WeChat had a lot of functions that were custom-built for Chinese social habits. For example, a friend request needs to be "approved" by the other party before a chat can take place; to add more friends (and hence make the app sticky), users can discover "people nearby" or find someone who is "shaking" their phone at the same time.

An executive who was working at WeChat India at that time told us, many years later, that he still vividly remembered the impossibility of teaching Indian users how to add friends. "People just do not get it," he said. "In comparison, whoever is in your contact list automatically becomes your WhatsApp connection."

In addition, the "shake" and "people nearby" features were often abused by ill-intentioned male users to harass female users. Many female users could not figure out how to turn off these features, so they simply stopped using WeChat altogether.

It is important to note that Chinese users did not have all the additional features at once. Instead, they grew with WeChat, with all the new features added incrementally, and thus the learning curve was much less steep.

Realizing that WeChat had hit a dead end in India, and not being willing to lose the market, Tencent led a funding round and invested in India's Hike messenger. However, by that time WhatsApp had gained a large customer base, and the network effect in the messaging space had strengthened WhatsApp's competitive advantage. It became impossible for Hike (or any other similar product) to shake WhatsApp's dominant position in India.

Product complexity is *not* the only cause of WeChat's failure outside mainland China. In Thailand and Taiwan, WeChat lost out to Line, a Japanese rival then owned by Korean company Naver (later sold to SoftBank). Users in both markets (especially Taiwan) had a lot of similarities to Chinese social habits. And Line was very similar to WeChat in many aspects.

It turned out that a key reason Line captured users in those two markets was the use of cute stickers, wildly popular in East and Southeast Asia. WeChat also had stickers, but they were far behind Line in terms of variety and cuteness.

People familiar with decision making at Tencent back then told us that a key struggle was that stickers were not a key feature in China, so the WeChat team could not figure out what to do with them. "Shall we allow anyone to release stickers? Shall we charge users to acquire stickers?" Some of these discussions were still taking place while Line quickly occupied the Thailand and Taiwan markets.

So, the question remains: Did Tencent, which had a market cap in the global top 10 list with more than a decade of tech experience, have the tech resources and capabilities to fix these product issues based on user feedback?

However, such fixes, which retrospectively were obvious actions to take, were not that straightforward back then, and required overcoming the challenges of organization, people, and leadership we discussed in Chapters 4, 5, and 6.

DIFFERENT APPROACHES TO EXPANSION: THE QUESTION OF HOW

So far we have discussed the *where, when,* and *what* questions about market expansion, but we have not yet addressed the question of *how.* There are multiple methods here, including building, joint ventures, and investment.

We put this consideration in the product chapter because we feel that this discussion is actually more related to product planning and operations than it is to organization and leadership. Of course, it is also closely related to those, with both being indispensable in determining the approach and making it work.

Build, Borrow, or Buy

Professors Laurence Capron and Will Mitchell suggested the expression "build, borrow, or buy"[2] (BBB—not related to President Biden's Covid recovery plan, Build Back Better).

BBB broadly describes three paths to enter new markets or pursue new growth initiatives (see Table 7.1). "Build" refers to companies establishing a new presence in the foreign market all by themselves, relying on their own

TABLE 7.1   Build, Borrow, and Buy in Overseas Expansion

	Build (Plant)	Borrow (Partner)	Buy (Purchase)
Active engagement	Greenfield ventures (ByteDance's Tiktok)	JV/Active alliance partners (Ant in Southeast Asia, JD)	Fully integrated acquisition (Alibaba and Lazada)
Passive engagement	Airborne entry (Baidu's early days in Brazil)	Passive investor (Sea group—Tencent investment)	Separately operated acquisition (Similar to passive investor)

resources and capabilities to self-plant new subsidiaries. On the other end of the spectrum is "buy," which refers to entering the foreign market by acquiring others' existing businesses in different countries. It is a relatively quick method to gain others' resources, capabilities, and local expertise through purchasing their assets and operations.

"Borrow" is somewhere in the middle—you do not fully rely on your own resources, nor do you solely depend on others. Instead, you try to partner with others to enter the new market, borrowing your strategic partners' resources and capabilities in the foreign market to achieve your purpose in the global market.

The other dimension—active and passive engagement—is from the perspective of managerial attention, suggesting the extent to which the executives in the headquarters spend their efforts on the entities they established in the foreign market, ranging from being substantially involved in the local business operations to minimum interference.

Thus, three paths and two types of engagement result in six potential expansion strategies: greenfield venture, airborne, joint venture (JV), passive investors, fully integrated acquisition, and separately operated acquisition. Tech companies venturing overseas may use a combination of these strategies, and they may also evolve over time, for instance, starting from JV and moving to fully integrated acquisition.

Build

Greenfield venture: A company setting up foreign businesses using its own resources and being actively involved in its management; examples are ByteDance's TikTok, or AliExpress.

Greenfield ventures have the benefits of full control over strategy and operation. You can set your own agenda, the pace of entry, and the region priority. You have the final word and full discretion on how much to invest, how to integrate the existing supply chain, how to hire, and how to manage the relationship between the headquarters and subsidiary.

However, the firm now bears the highest risk as a consequence of ownership. This model requires the firm to have the full team and all the capabilities needed to run the subsidiaries. In this case, organizational learning, experimentation and adjustment, and swift response to the market become super critical. For example, TikTok had to spend a lot of effort engaging with different regulators, especially in more conservative countries, on its content moderation.

Airborne entry: Building foreign market presence with a minimum involvement on the ground; this is typically adopted by pure internet models. Examples are some mobile gaming companies, or SheIn.

In this mode, the company does not set up operations in local markets; rather, they rely on the existing infrastructure to deploy their solutions into the foreign market. In traditional industries, this is similar to the foreign export whose purpose is to sell merchandise to overseas markets.

TANG BINSEN AND HIS GAMING BUSINESSES

The most popular online game in China since 2008 is *Happy Farm*, a multiplayer social game where the users play as the owner of a farm and grow a variety of vegetables and fruits. You can weed, use pesticides, water your plants, and harvest, or help your friends to do so, or even steal their plants. The *Happy Farm* game was promoted to more than 20 countries, acquiring 500 million overseas users, and is one of the most successful Chinese games venturing overseas. The man behind it is Tang Binshen.

In 2014, Tang's company launched a mobile massively multiplayer online strategy game, *Clash of Kings*, which quickly gained popularity. The game was downloaded more than 65 million times during its first year. It placed sixth on the best-seller list in North America.

Tang also launched web browser, search, navigation, and antivirus tools targeting emerging markets. What is common among all these businesses that Tang built was that he didn't really need anyone actually physically present in the target markets.

After making his money from gaming, Tang went into FMCG (fast-moving consumer goods) and created a blockbuster fizzy drink brand called Genki Forest in China. As of the end of 2021, Genki Forest was valued at US$15 billion. ▪

This model was fashionable before 2018 and is gradually fading. The reason is simple: more successful airborne companies realize that their business in destination markets is growing fast and that they will not be able to cope with things like customer services and partnerships without having a local presence.

For example, when SheIn encountered customs problems in Indonesia in mid-2021, without having its own local presence and depending fully on the service provider network it was not able to solve the problem on its own and subsequently decided to exit the market instead.

Nevertheless, this mode has its own benefits—low commitment and risk. The downside risk is minimal. It could be a viable approach for entrepreneurs to test business hypotheses and accumulate knowledge and expertise before making larger engagements in the market.

Borrow

Joint venture/active alliance partners: Pooling resources with one or more (often local) partners, and actively involved in their management. Examples are Ant Group's digital wallet joint ventures across Southeast Asia and JD's ecommerce joint ventures in Thailand and Indonesia.

In this case, local partners can contribute to the JV with its local knowledge, expertise, and social network to connect local business elites and regulators. This is also a very established method for large international consumer brands to enter emerging markets.

For instance, JD's ecommerce platforms with self-built warehousing and logistics systems in Thailand and Indonesia are joint ventures with Central Group and Provident Fund, respectively. Joint ventures allow JD to tap into resources, expertise, and relations of their local connection.

Ant Group adopted a similar approach in entering e-wallet businesses in South Asia and Southeast Asia (see Table 7.2). Its strategic partners vary from telecom companies to financial institutions to local business conglomerates. These strategic partners help Ant to comply with local regulations to penetrate local context in adopting mobile payment systems, among others.

TABLE 7.2 Ant Group's JV Partners in Its e-Wallet Business

Country	e-Wallet	JV Strategic Partner
Thailand	Ascend Money	CP Group (conglomerate)
Philippines	Mynt	Globe (telco)
Indonesia	DANA	Emtek (media conglomerate)
Malaysia	Touch 'n Go	CIMB Bank (financial)
Pakistan	EasyPaisa	Telenor Group (telco)

JVs help to reduce the risk of overseas exploration as a company works with local partners. However, JVs have their own challenges: they need to adhere to the goals of their partners, which are not always aligned with their own.

In the case of JD in Thailand, JD's objective is probably to develop ecommerce quickly to be competitive against other major platforms such as Shopee and Lazada, whereas Central Group is more concerned with acquiring ecommerce expertise as a defensive method to protect its massive offline retail business.

Almost all the joint ventures we have witnessed, from Ant-Emtek-Dana in Indonesia to Ping'an-Grab in Southeast Asia, have some degree of incentive misalignment.

In theory, regardless of expansion mode, the firm needs one or more local partners. While on paper many such partnerships point to obvious win-win situations, in reality many things could go wrong when interests or opinions misalign.

It is therefore critical to choose the right partner. Thorough due diligence is absolutely required before any decision is made. We have recently seen at least half a dozen cases where the Chinese tech firm is stuck with a local partner that does not align with them in strategy and style, but it reaches a point that too much is at stake to break the partnership.

Passive investor: A minority stakeholder with varying degrees of influence in direction. Examples are Tencent in Sea Group, and Alibaba in PayTM.

Investing sounds easy, as large Chinese tech companies typically have no shortage of cash and can invest at various stages in home-grown tech companies in destination markets.

Investing, however, means that the investing firm needs to position themselves correctly. We have seen a few (albeit rare) cases of large Chinese companies learning the market through investing, and subsequently deciding to enter on their own, crushing the home-grown companies they had invested in. We have also seen companies attempting to invest in all leading players in a particular sector, reducing the risk of concentrated bets but also the trust investees have in them.

While each company will have its own circumstances, perhaps Tencent's investment in Sea Group will be a case worth studying. As Sea's largest shareholder, Tencent gave the company substantial support in distributing popular games and in operational expertise, in addition to cash. At the same time, Tencent does not meddle with Sea's own operation and decision making, *at all*. Sea's *Freefire* game actually competes against Tencent's *Arena of Valor* in international markets, and Tencent seems to be perfectly fine with that.

The secret of Tencent's success is that it is more of a passive investor that does not directly interfere in operations. Meanwhile, Tencent is ready to share

TENCENT'S GROWTH STRATEGY

Tencent was a copycat in the early days. Based on its large number of users (fed by QQ and WeChat) and strong product management ability, it quickly copied a similar product, once Tencent found a good idea in the market, distributed it through its platform, and used its large customer base to squeeze the idea originator. For example, Tencent Weibo copied Sina Weibo, QQ whirlwind copied Thunder, Tencent's news friends network copied Renren.com . . . you name it. Despite earning a lot of money, Tencent didn't have a good reputation and was not respected. Copying was Tencent's style in its early stage.

But instead of copying and doing everything by itself, Tencent gradually became a mature investor. Its role is essentially an enabler of its investment targets. Tencent's investment in Sea and other investments in gaming businesses showed great success. ▪

knowledge, expertise, and even access to its consumer base. The relatively passive role gives Tencent's investees full discretion and strong incentives to develop the local market, and they indeed achieve a win-win situation.

In comparison, Alibaba and Ant's investees in Southeast Asia often complain about the group trying to exercise too much control, especially through back-end data systems. Alibaba and Ant executives will often force some changes on home-ground business founders to benefit Alibaba Group, at the cost of their own businesses, and thereafter often create tensions or even failures.

Buy

Acquisition and full integration: Purchasing/acquisition of an existing business. An example is Alibaba acquiring Lazada (Southeast Asia), Trendyol (Turkey), and Daraz (South Asia ex-India).

This approach has several advantages. The business is already established, with core market experience learned through practice. This saves the buyer a significant amount of time, reducing the early-stage failure risks and, crucially, the risk of missing their window of opportunity.

That said, the risk is when the buyer decides on a full integration. As is typical within a large tech conglomerate, the merger and acquisition (M&A) team is separate from the post-acquisition integration team, and coordination might not be fully smooth.

M-DAQ AND QUIXEY

In 2016, Ant Group invested tens of millions of dollars to acquire 40% of the shares of M-DAQ, a Singapore-based fintech company providing a platform for cross-border currency trading. Alibaba's platforms, including AliExpress and Tmall, had already been customers of M-DAQ before the investment.

After the investment, the Alibaba ecosystem became the main revenue resource of M-DAQ. Since the cooperation with Ant Financial and Alibaba Group's international e-commerce companies, M-DAQ's revenue has soared in the past few years, with annual sales surging from US$8.5 million in 2017 to US$31.4 million in 2018.

Due to M-DAQ's focus being entirely on serving Alibaba, it made Alibaba-related services account for more than 90% of the company's total revenue. At the same time, Ant Group is also constantly developing its own technology, which gradually shifts away from M-DAQ. The relationship between the two has become very tense. M-DAQ hopes to expand its business scope, get more autonomy, and explore financial technology opportunities. Ant Group may become a potential obstacle in this process.

A similar scenario happened to Quixey, a startup from California, which was a mobile search engine that scanned the major app stores and crawled blogs, review sites, forums, and social media sites to build a comprehensive picture of what an app can do—through reviews, word of mouth, and demos. Quixey received a $60 million investment led by Alibaba in 2015. However, in the following years, the two had serious disputes about technology, management, and so on, which caused Quixey's business to stagnate. In the end, Quixey closed in 2017.

Alibaba's investment is a double-edged sword for a startup. On the one hand, it can effectively help startups to grow through business synergies. For example, Ant Group successfully helped M-DAQ enter the ecommerce field, and Alibaba has high requirements for all its suppliers and partners, which also improved the technology and infrastructure of M-DAQ. On the other hand, it could potentially squeeze the potential growth of its investees.

Compared with Tencent, Alibaba Group has a stronger corporate culture, and is more inclined to control a company through investment, bind it with its huge business, and interfere with the company's operations to a certain extent.

Alibaba's acquisition of Lazada discussed earlier is a case in point. When the acquisition happened in 2016, Alibaba had neither managers with international ecommerce experience nor experience with coordinating a large business subsidiary outside China. A lot of time and energy were consumed during the process of fitting, which coincided with Shopee's meteoric rise. Baidu's acquisition of Peixe Urbano in Brazil and Tencent's acquisition of Ibibo in India, mentioned earlier, faced similar challenges.

Tencent's acquisition of iflix, a video-streaming player in Southeast Asia, however, followed a different trajectory. The Tencent WeTV team, which already had experience running streaming in Southeast Asia, took over quickly. The acquisition is recent (in 2020) and it is still too early to judge whether it is a success—but early signs show some promise.

A more certain example of success is Didi's acquisition of 99, a Brazilian ride-hailing company. This is partly due to the fact that ride hailing, as a business model, is much more straightforward than ecommerce, or content, in post-acquisition integration.

Acquisition and separate operation: Makes an acquisition but chooses not to actively engage in the operation of the acquired target. Examples are ByteDance's acquisition of Moonton Games, and Tencent's acquisition of Riot Games.

Firms may not be eager to integrate their acquired target into the parent companies for different reasons. One reason could be because of their strategic goal: Tencent's strategy is to allow the target firm's managers to have strong incentives and independence in decision making, and to allow internal competition.

Such acquisition, often, is strategic, allowing the buyer to own a piece that might be beneficial and critical for the future, and avoid the same piece from falling into the hands of competitors.

Indeed, for Chinese firms that are a newbie in the global business, waiting for some time until the market condition is mature is better than a harsh integration without a proper team and mutual understanding. As one executive once said, "When we become more mature and understand each other better, we could do things together."

Multiple Levers

While the previous discussion is laid out in a sequence from build to borrow to buy, there is no one mode that is always better than the other. Which mode is best depends on a firm's strategic goal and commitment, willingness to control, and risk tolerance, as well as circumstances and market reality.

In addition, firms can use multiple levers to achieve their goal of overseas expansion. For instance, ByteDance first used a mode of greenfield venture by launching TikTok specifically for the overseas market. It later acquired musical.ly and integrated it well into TikTok, which was a landmark in its growth trajectory.

Regardless of the expansion mode, when market opportunities emerge, operations actually start, and the clock is ticking, you need to make decisions and implement them.

As we can see, products are important, but they are just one element of the whole puzzle in the long journey of new overseas markets. To pace yourself well, make the right decisions, employ the right people, and leverage the right resources, organizing to deliver what you promise to the market, there is a lot to be done.

PUTTING THINGS TOGETHER: SPEED, SEQUENCE, AND RHYTHM

While the preceding sections highlight different dimensions of product, business decisions can be easily decomposed into small pieces and then put back together. In addition, what makes the overseas journey tougher for Chinese tech firms is how to pace the speed, sequence, and rhythm in their expansion.

In the seventh century BC, the Chinese military strategist Cao Gui famously said: 一鼓作气, 再而衰, 三而竭 (Sound the drum for the first time, the army is vigorous; second time, it becomes weak; third time, it becomes exhausted.)

Tech companies entering new markets also need to pace themselves so that they do not become exhausted. New markets have a lot of uncertainties and quite often the speed of business development is far from a company's ideal case. If not managed well, the team could very easily lose morale and disintegrate.

When a company or a venture becomes exhausted, good people will leave, and investors will lose patience. It takes extraordinary leadership to turn things around.

Therefore, the correct approach is to pace well, and grow with the market. SheIn, a Chinese cross-border ecommerce company, was for a long time seen as an underdog compared to more famous rivals JollyChic and Club Factory. Both latter companies were aggressive with their investment, while SheIn's management focused on building tech and data capabilities and a sophisticated supply chain. In 2020, when the two companies crumbled, SheIn more than doubled its annual sales to close to $10 billion.

To summarize, all these interrelated decisions pose great challenges to business leaders. However, compared to people and organization, which are a higher bar to overcome, we believe that Chinese tech firm leaders are more capable in dealing with product-related decisions, as many of them started their career as product managers, and understand the need to listen to the customers.

To conclude this chapter, we ask you to answer the following questions when you step into overseas territory:

- Is the timing right for the product you intend to launch in the specific market?
- Which stage is the market at, in infrastructure, consumer education, and so on, and how do you adapt your product accordingly?
- Do you have the people, organizational capabilities, and know-how to localize your product fast through iteration?
- Are your experienced product managers able to communicate seamlessly with local product managers who know the market?
- Is building yourself the best approach for this specific product? Will investment, joint venture, or other approaches make sense? How do you make such a decision, and what backup do you have?
- How do you keep your product agile and adaptable to the potential future changes in the market?
- How do you make sure you do not lose product momentum, and instead keep building on it?

NOTES

1. In 2021, Alibaba's Tmall Double 11 shopping festival achieved a new record of US$84.56 billion in gross merchandise volume.
2. L. Capron and W. Mitchell, *Build, Borrow, or Buy: Solving the Growth Dilemma* (Harvard Business Review Press, 2012).

PART THREE

Resteering the Wheel

CHAPTER EIGHT

The Inflection Point

There had been a lot of changes and so people are concerned. But everything happens for a reason. So, I would like to say, all for common prosperity. Even I would like to say, common prosperity is built in the genes of Meituan.

—*Wang Xing, founder of Meituan, 2021 Q2 earnings call*

THE YEAR 2021 WAS AN exceptional, and tough, one for large Chinese tech companies. The successive waves of regulatory actions not only sent most tech companies' share prices into a nosedive, but also triggered a nationwide soul-searching from tech leaders to investors to operation executives.

After years of running fast and intensive competitions, tech in China is destined to take a break, reflecting on its social value and resetting for the journey forward.

To understand what happened in 2021, we might need to start from the paddy fields in Indonesia.

In the early 1960s, American anthropologist Clifford Geertz studied rice farming in the Indonesian islands of Java and Bali in great detail and published a book called *Agricultural Involution: The Processes of Ecological Change in Indonesia.* In this book he coined the term "involution," which described the process in Java where internal and external pressures increased labor intensity in the paddies without increasing the output proportionally.

Or in simpler terms, everyone is forced to work extra hard for very little additional return.

What Geertz did not anticipate is that the term *involution* (内卷) would become the most used buzzword in China in 2021. People, especially the young, use this term to describe the circumstances they are in: having to work extra hard to survive in the competition, yet the extra work does not really bring extra benefits in the form of increased income or growth opportunities.

The 996 work culture, which people in the past two decades had embraced, is now resented. On December 29, 2020, a 23-year-old female Pinduoduo employee died suddenly after leaving work at 1:30 in the morning.

This triggered not only a big backlash against the large tech companies' work culture, but also soul searching for the youth. Unlike their predecessors in the previous decades, many of these young people realize that no matter how hard they work, they could not earn more income, afford an apartment in big cities, or get a promotion, even if they were working for the best names in tech.

What happened? After years of rapid development, tech has . . . slowed down. Obvious further growth opportunities are limited, while competition among companies as well as individuals has intensified. Gone are the days of natural growth, where you did not need to think too much, just work hard and work faster, and you would be naturally rewarded. Instead, involution has developed.

Another social phenomenon for the youth in 2021 was *lying flat* (躺平). This is basically an attitude where you live with the minimal resources, and you try not to compete with anyone for anything. On social media, many young people have said that lying flat resonates with their life attitude. Opportunities are so scarce, and competition is so fierce, why bother?

However, the truth is, there are growth opportunities. After two decades of building, the innovations in China's tech sector are actually cascading—each wave will create the infrastructure for the next wave. Figure 8.1 shows companies in each wave, from web to mobile internet and more recent development.

The problem is, there is an oversupply of talent and capital in the tech sector, which means many are not getting the returns they are expecting.

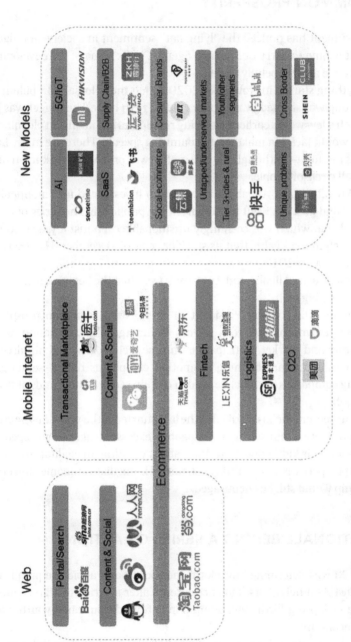

FIGURE 8.1 Companies in each wave
Source: Momentum Works.

COMMON PROSPERITY

The government has noticed the "lying flat" sentiment in society, and figured out that it is time to carry out its own agenda, to fix some of the development problems once and (hopefully) for all.

Everything started in November 2020, when the colossal $38 billion IPO of Ant Group was shelved by regulators at the 11th hour. The trigger was Jack Ma's speech a few weeks earlier, criticizing regulators for having an old mindset.

Why would Jack Ma make such a damning speech? There are many interpretations in China—although the real motive was probably only known to Ma and a small circle of people.

Quickly afterward, Alibaba itself was also investigated for "monopolistic practices," and Jack Ma went missing from the public eye. The series of events caused Alibaba, which was seemingly unstoppable even just a year before, to shed so much market value that its stock price as of December 31, 2021, was the same as in May 2017.

The issues with Alibaba and Ant Group were just the beginning of the turbulent year for big tech.

Events in 2021, where President Xi emphasized "common prosperity" and cracked down on some tech sectors including online education, unsettled investors and market participants. Chinese tech stocks shed hundreds of billions in value; money-burning promotion campaigns halted, sending big advertising companies into decline; and in certain sectors, thousands of people lost their jobs.

Various senior officials clarified in the last quarter of the year that the crackdowns were more about setting rules against excesses (e.g., parents spending too much on their kids' education) than about creating an egalitarian society. Common prosperity is an eventual goal, not an immediate outcome, and entrepreneurship would still be encouraged.

 ## RATIONALE BEHIND A SERIES OF ACTIONS

President Xi was determined to reform the society and avoid the pitfalls that China's East Asia neighbors had encountered after a period of high economic growth: a collapsing birth rate, aging population, disillusioned youth, and a stagnant economy.

Take the example of wiping out the whole online education industry: the objective is to free children from incessant competition (involution), but also to free parents from the high costs of raising children. President Xi wrote about the necessity for such educational reforms in his book *The Governance of China*, published much earlier, in 2014.

WHAT PRESIDENT XI WROTE IN 2014: EXCERPTS FROM *THE GOVERNANCE OF CHINA*

// The most prominent problem in education is that primary and secondary school students are exhausted and stressed. Some of the approaches that schools take are short-sighted and utilitarian, posing toxic levels of stress to students. What's more serious is that everyone knows that this situation is wrong, but they are walking along this road getting more and more trapped.

"Some off-campus training institutions violate the educational rules, speed up students' learning progress too fast, and carry out "exam-oriented" training, which not only increases the extracurricular burden of students and the financial burden of families, but also disrupts the regular school curriculum. The social response is strong. An industry of conscience cannot become a profit-seeking industry. We should have rules to manage the off-campus training institutions, so that off-campus education can return to the normal track." ▪

If you put this online education crackdown together with the three-child policy announced just a few weeks earlier, you would understand the thinking behind such policies. The online education industry as well as its employees became a necessary sacrifice for the greater good.

Apparently, *The Governance of China* became a top-selling book among investors and tech executives, who desperately tried to understand which other industries were seen as problematic.

Indeed, other departments of the state also acted. The State Administration for Market Regulation (SAMR), previously little known to the outside world, became an active player, investigating and fining players including Tencent, Alibaba, and Meituan for monopolistic wrongdoings.

On the antitrust front, the government has been quite busy too. Tencent Music was forced to relinquish exclusive distribution rights with music labels; Meituan was fined for forcing merchants to choose sides; and a series of investments and acquisitions from years earlier were ordered to be unbundled.

And in November 2021, the State Anti-Monopoly Bureau was established under SAMR to investigate and process antitrust cases. Friends familiar with the Chinese state establishment told us that the bureau, whose chief now holds an administrative rank of deputy minister, will have to work hard to prove their value at least in the few years to come.

Figure 8.2 lists the tech firms and sectors that have been recently influenced by the government.

Another particularly busy agency was the Cyberspace Administration of China (CAC). The agency, whose purview includes cyber- and data security, acted swiftly after Didi, China's ride-hailing giant, went for a US IPO in June without the consent of the Chinese government. Didi's data was deemed to be sensitive for national security.

Over the next half a year, the CAC launched a series of rules and regulations that essentially made US IPOs for Chinese consumer tech companies very difficult to enact.

While we can debate whether the series of actions are too brutal and that may have negative implications on firm growth, unemployment, and even cause economic slowdown, the following case suggests the market perhaps needs the government's intervention.

 ## WAR OF A THOUSAND GROUPS 2.0

As mentioned earlier in this chapter, as growth opportunities became limited, competition among big tech firms in China intensified. A key industry in late 2020 was community group buying (CGB). The business model here is simple: develop an agent network in the community that promotes products through WeChat, and consumers will pick up their orders from the agents. Pre-orders are aggregated, therefore inventory, wastage, and fulfillment cost per order is minimized.

Anyone who succeeds in community group buying ends up with a hyper-local, ultra-efficient fulfillment network on which they can theoretically sell anything.

Like the original group buying a decade ago, the CGB sector quickly descended into a messy war.

Alibaba, Meituan, Pinduoduo, JD.com, and Didi all joined the battle, with some of these players launching more than one service to diversify their risk. Figure 8.3 lists a few players of CGB in late 2020; most of them are gone today.

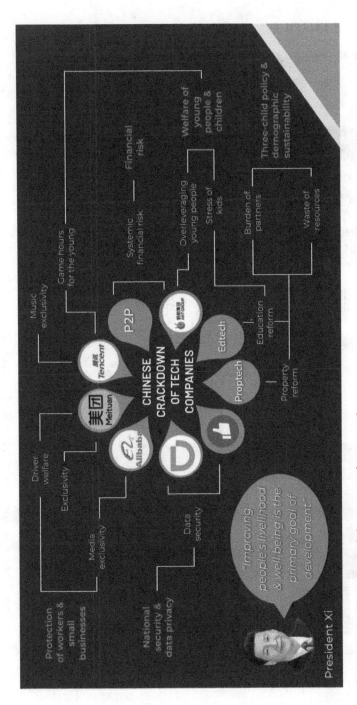

FIGURE 8.2 Chinese crackdown on tech companies
Source: Momentum Works.

FIGURE 8.3 Community group-buy players in late 2020

Source: https://m.news.leju.com/news-6740611648178142294.html.

Unfortunately, in this battle smaller players and startups will find it very hard to match the firepower of major tech firms. Even if they can execute better, it is hard to convince investors to put more money into fighting directly against the giants. By the end of 2021, among the community group buying startups only Xingsheng was still standing, to a certain extent thanks to its investor Tencent, which did not run its own community group buying platform.

Unlike the original group buying war, this time the regulators decided to step in early. This is for a number of reasons: first, the major players are now deploying a much larger amount of capital, sometimes even paying customers to take orders, disrupting the offline trade that millions of small businessmen depend on; second, the incessant competition, as with the 996 culture described earlier, has caused backlash in the society.

Chinese regulators intervened in March 2021, fining the top five players a total of RMB 6.5 billion (US$1 billion). Although the number was not significant to these big players, the message was clear: it is no longer acceptable to compete aggressively, without boundaries.

While regulators in the West have also aired concerns, and tried to rein in big techs, the Chinese methods have offered a quick and effective solution— albeit a double-edged sword.

We expect the sword to be used more often in the coming years, with its seemingly clean cuts, impact, and implications that will be long-lasting.

VENTURING OVERSEAS AS A NECESSITY, NOT A CHOICE

Under the macro environment change in China over the past year, we have seen many more Chinese companies and investors exploring global markets, establishing presences in Singapore, building distribution networks in the United States, and penetrating markets in Brazil and India.

For many, venturing overseas has become a necessity, one way (or the only way) to survive. Many companies are unprepared or underprepared—we have seen fintech companies approaching foreign partners and regulators even when they do not have any brochures or presentations in English.

The issues of leadership, people, organization, and product are happening again and again, in different permutations and nuances but essentially following the same macro themes.

However, do not underestimate the survival instinct of Chinese businesses. Many will fail in their journey overseas, but collectively they learn from each other, build upon each other, and some will eventually stand out to shine.

CHAPTER NINE

The Global Chinese Community Fills the Gaps

WHILE CHINESE COMPANIES VENTURING overseas often encounter a lot of challenges, especially when they already have a major business back home in China, the global Chinese entrepreneurial community is taking some of the learnings from China, internalizing these experiences very well, and leveraging their strong understanding in overseas markets to build successful businesses.

The two companies that emerged in Southeast Asia and quickly evolved into global giants, Sea Group and J&T Express, are perhaps the best examples of this group. Founders of both companies originally came from China, but had lived, studied, and/or worked in Southeast Asia for years.

Some might argue that US-based companies such as Zoom and DoorDash fall into this category too. For example, DoorDash's founder immigrated to the United States from the Chinese city of Nanjing at a young age. As a result, the company has much better access to Meituan's experiences in food delivery, as well as a better capacity to study, understand, and adapt some of these learnings.

There are more: Stori, a neobank in Mexico; HungryPanda, a London-based food delivery platform; Flash Express, a Thai ecommerce logistics company; Wiz, a Singapore-based voice AI startup; the list goes on.

"This group of entrepreneurs originally from Mainland China is creating impact in consumer tech in a similar way to how the Taiwanese dominated the semiconductor industry," one prominent consumer tech investor quipped in his interview with us for this book. His reference is clear: the CEOs of both AMD and Nvidia, two of the most prominent chip designers, were both born in Tainan, an ancient city in Taiwan.

Compared to the "Chuhai" (venturing overseas) group of Chinese entrepreneurs, who are leaving China to look for tech opportunities elsewhere, the global Chinese community has a distinctive advantage: they know the local markets really well—not only how to localize a product, but also how to build a hybrid organization that combines the best of proven experiences from China and understanding local markets.

"Shopee manages its Brazilian workforce much more effectively than all the Chinese tech companies I have seen," a São Paulo-based Chinese tech executive said, somewhat mistakenly thinking that Shopee, the ecommerce arm of Sea Group, is a Chinese company.

Years of growth of tech in China in the 2000s and 2010s also created a large cohort of ethnic-Chinese fund managers who have studied and understood these companies and business models, as well as the evolving competitive dynamics, very well. Some of these fund managers have been advising their global investees in acquiring not only strategies, but also operational structures and even key executives from their counterparts in China.

The case of Sea Group is particularly intriguing here. The company, which originated as Garena, a game distributor in Southeast Asia, has morphed into a consumer tech conglomerate covering gaming, ecommerce, and digital financial services (Figure 9.1). Its footprint has also expanded globally, especially during the COVID-19 pandemic.

Between Q4 2019 and Q4 2021, the company's share price rose 10 times, before crashing down together with much of the emerging tech as this book was written, in the beginning of 2021.

Despite its size and presence, the company has been spectacularly quiet in public, leaving many business media outlets branding it as "secretive." Its annual reports are also quite barren compared to many of its peers'—not much chest beating or management grandiosity, just boring numbers and charts.

Yet on the ground in different markets it operates in, partners and competitors tell a very different story: the company, especially its ecommerce arm Shopee, is very aggressive in both expansion as well as competitive tactics.

Investors are divided. "We have not seen this level of money burning for a while," one hedge fund manager said, after crunching some numbers about Shopee's newly launched food-delivery services in Indonesia, Malaysia, and

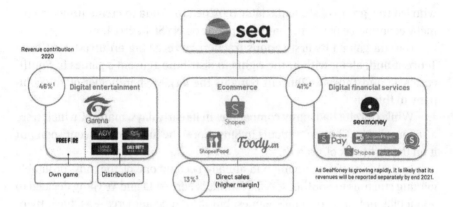

FIGURE 9.1 Organizational structure of Sea Group

[1] Primarily by selling in-game items to game players.
[2] Ecommerce and other services: Revenue from ecommerce marketplace, digital financial services, and other services on the platform.
[3] Sales of product: Revenue from product own and sold by Shopee on its platform.

Source: Sea Group 2021 annual report. Momentum Works research and insights. © Momentum Works.

Thailand in 2021, against very established players including Nasdaq-listed Grab and Germany's DAX-member DeliveryHero. You heard it right, that comment was made about Southeast Asia, where since 2016 billions after billions of dollars were invested in large consumer tech companies.

We have already mentioned some organization and product aspects of Shopee's competition against Alibaba's Lazada in Chapters 6 and 7, respectively. However, we think the story of Sea Group as a whole deserves a chapter on its own—it is, in our opinion, a successful adaptation of Chinese tech strategies and experiences in multiple global markets.

There is a lot we can learn from this experience, as well as Sea Group's competition against a whole suite of Chinese and non-Chinese players, including not only the previously mentioned Grab, DeliveryHero, and Alibaba, but also Amazon, TikTok, Poland's Allegro, Walmart's Indian affiliate Flipkart, Latin America's MercadoLibre, and potentially Brazil's neobank darling Nubank.

 ## SEA GROUP AND SHOPEE

Sea Group, formally known as Garena, was founded in Singapore in 2009, but it traces its roots back to a gaming platform built by some Chinese students at Singapore's Nanyang Technological University in the mid-2000s. Leo Chen, who led this group of students, later returned to China to create Jumei, a cosmetic ecommerce platform that went public on NYSE in 2014.

Garena gained its first serious traction by receiving an investment from Tencent and, along with it, the rights to distribute Tencent's games in Southeast Asia and Taiwan. (Tencent remains the largest shareholder of the company to this day.)

While Garena had many competitors in its early days, many of which were also founded by Chinese students in Singapore, the fuel from Tencent soon put it on the fast track of growth.

In 2014, as the company had already become one of the most profitable gaming startups in Southeast Asia, founders Forrest Li and Ye Gang decided to go mobile, not only in mobile games, but also in ecommerce—as, back then, the fast proliferation of smartphones in China meant that millions of consumers could have direct access to brands and products through ecommerce platforms. Would Southeast Asians, who were starting to adopt smartphones as well, eventually find ecommerce appealing?

To do this, they tapped Chris Feng, a former McKinsey consultant who by then had spent a number of years with Rocket Internet—a German venture group that had built multiple successful internet companies in markets outside China and the United States. At that time, Feng was a managing director of Lazada, Rocket Internet's ecommerce venture in Southeast Asia that was to be acquired by Alibaba two years later.

At Lazada, Feng had a reputation of being both very smart and "incredibly direct"—a rare attribute among Southeast Asia's often high-context, indirect cultural backgrounds.

The first version of Shopee, released in 2015, resembled Carousell, a popular mobile app for trading used items in Singapore. Media likened Shopee to Carousell, and anticipated that Shopee would face the same challenges Carousell faced, such as uncooperative buyers, quality issues of listed items, and the platform's inability to monetize.

In a December 2015 interview with *Vulcan Post*, Feng said that Shopee's vision is very different from that of Carousell:

> Carousell was very lucky in the sense that they started early in the Singapore market. It's good that they educate the market, ultimately it's like a welcome present. I don't see them as a competitor long term—they will be niche players, second-hand market, for a very long time, and it's okay to be like that. . . . In some sense we do look somehow similar, but I think we'll evolve very differently as time goes.

When Sea Group's market cap hit the historical high of US$200 billion in October 2021, Carousell finally reached the unicorn valuation of US$1 billion. With the major market still in Singapore and revenue of less than US$50 million, Carousell indeed evolved very differently, as Feng said in 2015.

The first commercial success of Shopee, unexpectedly to many observers, was in Taiwan, a prosperous, Chinese-speaking society of 23 million people. Taiwan was not an alien market to Shopee—Garena had been running a gaming business there for a couple of years, and Feng had personally led one of Rocket Internet's ecommerce launches in Taiwan a few years earlier.

The timing was perfect for Shopee. While consumption power and infrastructure development in Taiwan made it a perfect market for ecommerce growth, the main competitors of Shopee were Yahoo! and local directory PCHome. Both companies were from the desktop internet era and neither had invested in the mobile experience.

Another potentially big competitor, Alibaba's Taobao, had shown some good promise in the Taiwan market. However, politics played in Shopee's favor. After a pro-independence candidate won the presidential election in January 2016, it became very difficult for a mainland Chinese tech company to operate at large scale in Taiwan—it would be bad optics in Taiwan, where the authorities would regard Mainland China as aggressors, and in China, where the government regards Taiwan as a rebel province to be eventually reunified with the mainland.

To illustrate how much Shopee had learned from Chinese tech companies, look no further than the Chinese name it adopted in the Taiwan market—Xiapi (蝦皮, or dried shrimp). To put things in perspective, Alibaba's B2C ecommerce platform is called Tmall (天猫, literally Heavenly Cat in Chinese); financial services affiliate Ant Financial (later renamed as Ant Group); logistics platform Cainiao (菜鸟, or newbie bird, in Chinese); and omnichannel grocer Hema (盒马, or hippo).

The success for Shopee in Taiwan was swift. Sources we spoke to told us that by the end of 2016, Taiwan already made up 90% of Shopee's total volume. To reconcile Taiwan's importance in its portfolio, when Sea Group went for an IPO in 2017 it coined the term "Greater Southeast Asia," which also included Taiwan.

In fact, by 2017, Shopee had already made large entries into all major markets in Southeast Asia, and claimed to have overtaken Lazada in gross merchandise volume (GMV) terms by the end of the year. However, investors were also worried about the company's tactics: the extensive use of free shipping to grow the number of orders and customers. The company's IPO in 2017 was partially a reflection of the difficulty it encountered trying to raise money from the private equity market. The share price also barely moved in the full year after the IPO.

On the ground, however, Shopee was consistently growing its ecommerce penetration across Southeast Asia. Free shipping kept the wheel rolling—the company also built a large cross-border sourcing operation, onboarding and servicing Chinese sellers to provide additional selection of value-for-money goods into Southeast Asia. In 2019, Shopee overtook Tokopedia to become the largest ecommerce platform in Indonesia in GMV terms.

The company, which had already played with payment services in the early days of Garena, also acquired necessary licenses to roll out mobile and digital payment across the region. Toward the end of 2019, ShopeePay stickers started emerging on offline merchant stores—reminiscent of Alipay's offline push in China in the mid-2010s.

Investor confidence was lifted also after seeing the success of *Free Fire*, a multiplayer online battle arena (MOBA) game that Garena developed and launched in 2017. Not only did it generate large cash flows from almost across the world, but Sea Group could also invest this cash flow into growing its ecommerce operations.

The COVID-19 pandemic, which started in early 2020, provided a huge boost to all business lines of Sea Group. Under extensive social restrictions, people around the world shopped more, avoided contact while making payments, and played mobile games for long hours.

THE CHINESE TECH ZOO

An interesting fact about tech in China is the extensive use of animal names or images in brands. Alibaba's portfolio of companies is often affectionately called the "Alibaba Zoo." All of Alibaba's 29 businesses have an associated animal name or logos, such as music platform Xiami (shrimp), secondhand platform Xianyu (translated as "leisure fish" with a yellow fish logo), and logistics service Fenniao (hummingbird). See Figure 9.2.

Aside from Alibaba's Tmall—cat, other ecommerce platforms in China also used animals: JD—dog; Suning.com—lion; and Gome—tiger. See Figure 9.3.

FIGURE 9.2 The Alibaba Zoo

(Continued)

(Continued)

FIGURE 9.3 Other ecommerce animals

Tencent started as a penguin, and Meituan is a kangaroo, shown in Figure 9.4.

FIGURE 9.4 Tencent's penguin and Meituan's kangaroo

And there are many other more in various sectors, such as in travel sectors: Ctrip—dolphin, Qunar.com—camel, Feizhu—piggy, Tuniu.com—cow, eLong—dragon, mafengwo—wasp, Tongcheng—clownfish, and Lvmama.com—donkey, shown in Figure 9.5.

FIGURE 9.5 Other sectors' animal logos

Similarly, in live-streaming sectors, we have seen Douyu.com—shark, Panda TV—panda, and Inke—owl.

In addition to animals, company names and logos are also expanded to plant-based images, as shown in Figure 9.6. We can see a lily, tomato, peapod, pear, mushroom, grapefruit, pepper, nut, litchi, watermelon, even melon seeds.

FIGURE 9.6 Plant-based logos

There are different theories of why this came about. One argument is that animal names and images are easier to remember and circulate among the general public. The other is that because early entrepreneurs in China were developers, they were already familiar with using animals (Firefox—fox, Linux—penguin, GitHub—cat), shown in Figure 9.7.

FIGURE 9.7 Animal logos familiar to early entrepreneurs

Share prices surged, and the company used this to step on the accelerator:

- Shopee launched across five major markets in Latin America, Poland, France, and Spain in Europe, as well as the famously tough to crack market of India.
- It has also invested heavily in logistics infrastructure through Shopee Express, in an attempt to own fulfillment.
- It added more features, including live streaming, to the platform.

- In 2020, the company also formalized its digital financial services offerings into a subsidiary called SeaMoney, offering not only payment, but also buy-now-pay-later, cash lending, insurance, and other financial services.
- Over the past two years, Sea Group obtained a digital banking license in Singapore, and acquired banks or banking licenses in Indonesia and the Philippines.
- In 2021, Sea Group rebranded Foody Now, a food-delivery platform in Vietnam, as ShopeeFood—and similar services were launched and aggressively grown in Indonesia, Malaysia, and Thailand.
- In the same year, Sea Group also acquired the Hong Kong–based investment firm Composite Capital and rebranded it as Sea Capital.

In September 2021, Sea raised more than US$6 billion in additional equity issues and convertibles at US$318 per share. According to a person familiar with the process, the sentiment was so bullish that the entire lot was subscribed within two days.

To a question from one of the investors at a roadshow about why Sea needs that much cash, the head of Sea's investor relations answered along the lines of: The market opportunity is so big, our competitors are so weak, and we know how to deliver growth better than you do, so why shouldn't we?

Investors, and even some senior executives, however, were divided. While many enjoyed the hyperspeed of growth, some felt the company was over-stretched by doing too many things at the same time.

With sales of *Free Fire* slowing down as consumers across the world return to the streets, and no new hit game in Garena's pipeline, Shopee will need to continue to tap into the capital market to fund its growth.

When capital market sentiment changed, the whole world came crashing down. Like the good news, bad news also came in quick succession:

- In January 2022, Tencent divested 2.6% of its equity interest in Sea Group. Although Tencent still remains the largest shareholder—and a likely cause of the divestiture is SEA trying to reduce the perceived Chinese control in the increasingly complicated geopolitical environment—market confidence is nonetheless impacted.
- In February 2022, India banned *Free Fire*, apparently viewing it as a Chinese app. While India contributed very minimally to Garena's overall revenue, it was the largest market in terms of user base, and the ban raised doubts as to whether Shopee, which had already built a base in the country, would come next.

■ In early March 2022, Shopee shut down its France operations—less than four months after the launch. Suddenly, the belief that Shopee could keep expanding and beating rivals left right and center is all but shattered.

All this happened against the major backdrop of inflation in the United States, the expectation of tightening monetary policy, and the war in Ukraine, all of which already sent growth stocks into a nosedive. Sea's major emerging market competitors—Grab, Alibaba, Mercado Libre, Allegro, and DeliveryHero—all saw slashing market capitalizations.

While Sea has seen many uncertainties, challenges, and very difficult situations in the past, this time is different. The company is already massive, with more than 30,000 employees, operations in more than a dozen countries, and much higher expectations (many new employees' compensation package in shares and or options is based on a US$300+ price). The challenges to the leadership, organization, and people are at a much higher level.

Let's look at these aspects in slightly more detail—you will understand the parallels, but also the differences between Sea and many of the Chinese tech giants it took reference from.

LEADERSHIP AND STRATEGY

A common question when investors and analysts started covering Sea was: "Would Chris Feng one day leave the company?"

This question is an indication of Feng's critical role in building and steering the company forward.

As a nonfounder of the Group, Feng's shareholding is limited compared to his role. People should see no reason why he should leave to pursue other things. Ultimately, where else would he get such an outsized role, build businesses at such a scale, and have such trust from the founders?

Feng, who started (and still helms) Shopee, has subsequently assumed the CEO role for SeaMoney, the digital financial services unit whose business is closely related to Shopee. He is a strategic thinker but also very hands-on and detail-oriented—many key executives report to him directly rather than going through layers of corporate hierarchy. He personally made decisions, including how shopping festival ad campaigns should be run. Decisions are usually made very quickly.

With the increased complexity of the business, it is hard to believe that Feng is not overworked. Even when many very competent and experienced tech executives have joined his ranks, he would still need the mental space to deal with the myriad of decisions needed for the group.

The founders, Forrest Li and Ye Gang, understand this very well, and they have created an environment where Feng could focus on building the business. Anything to do with investor relations, government relations, licensing, and social responsibility has been taken care of by the Group, to clear the way for Feng's business operations in both ecommerce and digital financial services.

That arrangement has worked out well. In Q4 2021, Feng was appointed president of the Group, a recognition of his importance but also a signal to the market that he is deeply valued (and will not leave the group any time soon).

Feng's upbringing perhaps bears some resemblance to that of JD's founder, Richard Liu. Both were born in the countryside in the poorer, northern part of China's Jiangsu province—their hometowns were a bit more than a one-hour drive apart. The divergence in paths came when Feng obtained a scholarship to study in Singapore, where he acquired a degree in computer science at the National University of Singapore.

After spending seven years at McKinsey, which took him to places as far as Germany and Brazil on client projects, Feng joined Rocket Internet's ecommerce team for Asia. The team eventually evolved into two separate ventures: general ecommerce platform Lazada and fashion ecommerce platform Zalora.

Feng was initially focused on the launching and overseeing of procurement of Zalora; he later moved on to Lazada to oversee assortment, purchasing, and direct sourcing across all categories and for all countries.

Back in 2012–2014, Rocket Internet was most willing to spend money across emerging markets including Southeast Asia, which captured much of the region's restive tech and internet talent. One of the authors of this book was also working with Rocket Internet for two years during that period and vividly remembers the thrill and adrenaline of venture building. In 2011, Nadiem Makarin was also tapped from McKinsey to lead part of Zalora. He later went on to found Gojek, which became Indonesia's most valued tech company, covering mobility, food delivery, and digital financial services, and a key competitor of Shopee in Indonesia.

Feng spent two years with Rocket, and decided to move on to Garena to build Garena Mobile. He took some of his old McKinsey and Lazada lieutenants with him—some of whom are still working very closely with him to this day.

Despite the fact that Shopee was growing milestone after milestone, Feng was actually quite focused. An avid reader of Chinese history (as are most of the Chinese tech entrepreneurs we mentioned in earlier chapters), he understood the strategic challenges of overstretching, but also the balance between fast growth and strategic depth.

Crossing the Border (from China)

In managing Shopee's growth, Feng saw the advantage of the supply chain in China, which manufactured millions of different types of consumer products that power the ecommerce selection. A priority Shopee had in earlier days was actually merchandise sourcing. To do that, it built a large seller team in China, and worked with logistics and payment partners.

The best part: many sellers in China had already been selling across the border when Shopee approached them, on platforms including Amazon, AliExpress, Allegro (all three are Shopee competitors in Poland, interestingly), and Lazada. They do not need education, but they are often underserved—lacking insight, guidance, and sometimes on-time settlement of payments. Shopee can exactly tap into these areas and offer a slightly better experience to these merchants, saving the cost of having to educate them.

The wide selection of goods from China- (and, later on, Korea-) based cross-border merchants was often the first wave of attraction for consumers in new markets Shopee launched into.

Free Shipping

Cross-border sellers with a wide variety of consumer product choices, coupled with free shipping, discounts, and flash sale campaigns—all proven tactics in China—often quickly turn the Shopee flywheel, and give it a good foothold in the new market.

Free shipping is the simplest and most powerful marketing message to consumers. Taobao, in its early days, had famous "free shipping for Jiangsu-Zhejiang-Shanghai provinces," which proved powerful enough to generate large ecommerce adoption across the Yangtze River delta—the three provinces are among the richest in China, with a combined population of 174 million.

Free shipping is, however, costly, basically subsidizing every single order placed by consumers. Many investors and competitors were dismissive about Shopee's strategy early on precisely because they saw free shipping as nonsustainable.

Feng and his team stuck with it—they knew very well that free shipping offered them the shortest path to having millions of consumers on the platform. These consumers are then channeled to attracting and rewarding the best sellers. The improved offers from sellers in turn attract more consumers, and the increased volume reduces costs in operations, logistics, and payment, making the whole platform more competitive.

In a way, this is no different in spirit than Amazon's flywheel (Figure 9.8).

FIGURE 9.8 Amazon's and Shopee's strategies
Source: © Momentum Works.
* GRATIS ONGKIR means "Free Shipping" in Indonesian language.

Women Are the Most Important Customers

Shopee also recognized that to win the ecommerce game, young and female users are vital. Many of its gamified features, selection of assortments, and promotions target such a demographic group. It is also a well-known strategy across different ecommerce participants in China. Among Shopee's largest categories are home and living, female fashion, and beauty and cosmetics, which account for more than one third of total sales.

These three categories have several things in common: female decision makers, high margins, and significant supply chain advantages from China. These are all key decisions in Shopee's product strategy—what to offer, whom to target, and how to operate and grow.

Traffic → Data → Monetization

In practice, Shopee's strategy is more similar to Taobao's, where massive, low-value customers and orders are prioritized compared to more premium ones.

This strategy was puzzling to many investors, analysts, and competitors, but fairly familiar to anyone who has closely watched or participated in consumer tech growth in China.

"To understand all these years of Chinese consumer tech growth, you need to first understand this word: 流量 (consumer traffic)," one billionaire exited founder once told us.

Underpinning this strategy is that a large platform needs to start and dominate high-frequency, low-value transactions to capture and retain users, and then build high-value, low-frequency transactions on top to monetize. "The reverse has never worked in China," the same founder told us.

Alibaba started its consumer ecommerce with the low-value, high-frequency Taobao for massive volume, and then spun off Tmall as a separate brand for more premium, branded purchases to make a profit. Similarly, Meituan used its low-value and low-margin but high-frequency food-delivery business to achieve volume and profit from its high-value, lower-frequency travel and other businesses. Other successful large consumer tech platforms in China all used similar strategies.

The hierarchy of transaction cases in descending order in the conversion funnel (from the highest frequency to the lowest) is also applied in other domains of business, such as social and content (e.g., WeChat, TikTok) and payment (e.g., Alipay), in addition to ecommerce (e.g., Taobao), ride hailing, and food delivery.

Compared to many of its rivals in various markets, Shopee has applied this strategy very consistently, much to the bewilderment of its competitors. Lazada executives lamented in 2017 that Shopee customers were bargain hunters not worth pursuing; Latin America's MercadoLibre had the same stance in 2021; so do ShopeeFood's main competitors in Southeast Asia in 2022.

Why Do Competitors Not React?

One question that often puzzled investors and analysts who have followed China tech closely was "This all seems like common sense to us, why do Shopee's competitors not act or react accordingly to arrest Shopee's growth?"

First, this strategy is not necessarily common sense to everyone. There is still debate over whether success in China is country-specific (ultimately China is unique in many aspects) or widely applicable in other emerging markets too. Executives from Shopee competitors we spoke to are often dismissive of the strategy.

There are some reasons why the same strategy does not work everywhere every time. While consumers across Southeast Asia and in Brazil and Poland responded to these growth tactics well, consumers in France showed an obvious behavior of . . . indifference. Shopee could not get them to act on campaigns or discounts that work in other markets, resulting in very high effective customer acquisition costs—a key reason for its withdrawal from France in March 2022, less than four months after entry.

However, by 2021 it was clear to everyone that Shopee's strategy was winning in most markets in Southeast Asia. We somehow feel that the competitors, while now taking Shopee seriously enough, are still not entirely convinced about Shopee's strategy, or decisive enough to respond to it.

One example is Carousell (mentioned earlier in this chapter), which has not exactly been clear about whether they want to focus on just classifieds with advertising or close the loop by capturing transactions with a commission charged. Tokopedia, Shopee's largest competitor in Indonesia, initially struggled with Shopee's assortment of cross-border selection, and debated internally for a long time whether they should launch across the border as well. As a champion for enabling Indonesian SMEs, it was quite difficult for them to open up to Chinese sellers; however, the lack of a clear decision in this and many other areas allowed Shopee to eventually overtake Carousell to become the largest marketplace in Indonesia.

However, how about Alibaba affiliates Lazada and AliExpress, which operate in the same markets as Shopee? Taobao grew exactly with this strategy, so are Alibaba's other affiliates in China? Surely they understand and appreciate that?

Well, they do. However, for them the problem is much more nuanced. As mentioned in earlier chapters, organizational issues and decision-making mechanisms in Lazada (and AliExpress too) are often more complicated, which prevents consistent execution even if they are fully aware of what Shopee is up to, and have better technology, operational experience, and expertise to counter that.

Another challenge facing Alibaba companies is that they are too familiar with the ecommerce evolution and what advanced stages look like. That led to Lazada to focus a lot on brands earlier on, limiting itself to a premium position while Shopee was building volume. As mentioned earlier, it is much easier to launch into low-frequency, high-value transaction cases from high-frequency, low-value ones than the other way around. Lazada's attempts to compete against Shopee through free shipping falls exactly into this predicament.

Late-Mover Advantage

Another key advantage of Shopee is against the common perception of the first-mover advantage that is often discussed in highly competitive markets. However, Shopee, in Southeast Asia at least, has largely and almost deliberately chosen to be a late mover, and turning that into an advantage.

In 2016 when Shopee stepped on the pedal in Southeast Asia, Lazada had been around for four years; when ShopeePay started aggressive acquisition in Indonesia at the end of 2019, at least three other mobile wallets—Ant Group–backed Dana, Grab and Tokopedia–backed OVO, and Gojek's GoPay—had been fighting an intensive subsidy war for many months; Gojek and Grab's food delivery business had formed a duopoly when ShopeeFood entered Indonesia in 2021; not to mention the challenge to Allegro in Poland and MercadoLibre in Latin America.

How did Shopee do all that?

First, while time to market is critically important, the very notion of first-mover advantage needs to be decomposed. In the markets Shopee has chosen to enter to challenge the earlier movers, we often see a situation where the earlier movers have spent money, time, and effort educating the market, be it sellers, buyers, ecosystem partners, or their own employees; however, the market is yet to be fully penetrated. In other words, there is significant market potential that has yet to be fully captured, and the earlier movers have not yet reached the stage where they are able to consolidate the market leadership and build a moat.

Enter Shopee, with its decisive tactics: they can onboard sellers and customers who are already educated; poach the executives of competitors who are already well trained; and plug into the infrastructure that has already been built. In all these areas you just need to be better than the competition, and by observing closely the mistakes and weaknesses of competition, it is not that hard to figure out how to avoid them.

Then supercharge. This strategy has worked very well for Shopee in many business areas in many countries in Southeast Asia. And it allowed them a continuous momentum of growth while competitors had often exhausted themselves through all the market education.

As we mentioned in Chapter 7, Product, we highlight again the risk of first mover becoming first curse. The late mover, if executed well, has a huge advantage, as Duan Yongping, a famous Chinese tech entrepreneur, articulated clearly in Chapter 2.

Leadership on Timing and Prioritization

Underpinning the late-mover advantage is the leadership's, notably Feng's, clear sense of timing and prioritization. As a big organization with multiple business lines, there are many things that Shopee could do, and should do.

However, the resources and people's mental space are always limited. What you could do does not mean you should do it; what you should do does not mean you should do it now. Many companies in Southeast Asia, China, and beyond fumbled when they focused on the right things at the wrong time.

We heard from Shopee insiders that when his subordinates present a plan to resolve a problem or build a new feature, Feng's response is often "Is this the right time to do it?"

"He can always parse complex issues and determine whether they should be prioritized or left to later stages," an ex-Shopee executive we interviewed told us.

If you follow the product logic of Shopee, you will know: instead of building everything at once, Shopee has built capacities in many areas: payment, digital lending, food delivery, fresh groceries, Shopee Express logistics, and so on, but only stepped on the pedal for each of them when the timing was right. That was also the reason it could demonstrate almost sustained growth in each of the quarterly financial reports it released.

Another issue is the perception of people. At various stages of Shopee's, or Sea Group's in general, development, many employees will tell you that the company was problematic and that much of their advice was ignored. Upon closely watching Shopee's growth for the past five years, we feel that much improvement advice was ignored because, while valid, it should not have been the focus of the group at that particular point of time. The investment of leadership's time and mental space, as well as the group's money and resources, into certain areas will generate a much better return over both the short and long term.

This, although clear to the leadership, might not be that apparent to employees at specific posts. Those who understand the logic tend to stick with the company, while those who do not often end up frustrated and leaving.

PEOPLE, ORGANIZATION, AND PRODUCT

In addition to the preceding discussion of leadership and strategy, we further apply the POP-Leadership framework to zoom in people, organization, and product that collectively painted a picture for Sea Group and Shopee's historic growth and current challenges.

The Early Days

Pretty similar to Alibaba and other major tech successes in China we mentioned earlier, the early employees of Garena were not necessarily the best in the market. Facing an uncertain future at Garena, candidates who had a choice would join banks, consulting companies, and even semiconductor companies. Many of these earlier employees of Garena were Chinese graduates in Southeast Asia who were either passionate about (or addicted to) gaming, or did not speak English well enough to pass bank or consulting interviews.

Similar to Alibaba but probably less dramatically, the early employees also had their stories of when the company was small (when the whole team could be fit into a junk in Singapore River for team building), fun, and sometimes very stressful (when the cofounder knocks at your home's door to drag you out to fix a server problem). Many left the company for better opportunities earlier on, a decision some have come to regret deeply years down the road.

As Garena became increasingly profitable, the talent it was able to attract also improved. When the company first opened its new headquarters in Singapore's One North district, designated by the government as a zone for tech companies, it included unlimited snacks, napping pods, and a resident masseuse. This had been a tactic commonly used by Chinese tech companies to attract and motivate engineering talent who might work 14–16 hours a day. At that time in Singapore, only American tech companies such as Apple and Google offered gourmet food, and their talent in the region was hardly any tech.

Shopee's Organization and People

Shopee brought in a culture of its own, deeply influenced by Chris Feng. Communications were straightforward, decisions were fast, and key executives stressed over facts rather than presentation. That has allowed the company to be efficient, but also less attractive to experienced corporate executives, who were used to a totally different set of expectations and communication protocols.

While the chain of command seems to be very clear, the company seems to be conscious that the expectations of senior executives are very different from those of junior operatives. Shopee operates in different Southeast Asian markets with very different cultures and dynamics, and at the ground level it has to rely on local operatives to succeed. This means that it can only tap into the same talent pool as its competitors.

To win and to succeed, it needs to differentiate at the senior level, with competent and driven executives, to stay agile and adaptable to local market realities, while tapping effectively into the strategies and expertise from the regional team.

For a long time, Shopee did not assign the role of country managers; rather, Chris and his top lieutenant, Terence Pang, would spend a lot of time personally in the key markets, leading the team and growing the operations. Senior executives who are leading the cross-border team in China are asked to spend time in destination markets, so that they will gain a firsthand sense of consumers for their decision making as well.

At the same time, a group of young executives rose up through the ranks to increasingly assume more responsibilities. For example, Christin Djuarto, an Indonesian Chinese who had graduated from Singapore's Nanyang Techno-logical University and spent three years with Garena before moving to Shopee, became executive director and took on leadership responsibilities in Shopee's largest market.

Long-term lieutenants are also given the opportunity to develop new mar-kets. For example, Jianghong Liu, who previously worked as head of cross bor-der based in Shenzhen, China, was assigned to lead the operations in Mexico; Pine Kyaw, who previously led Shopee Vietnam, was made executive director to lead its Brazil business, among a group of other executives sent from Asia.

When Shopee launched digital financial services, including ShopeePay as well as consumer credit, it struggled for a while to get the right senior executives. China's crackdown of fintech and especially lending since 2019 has provided a golden opportunity—a group of experienced executives who found China's market no longer welcome ended up at Shopee, which they consider not too alien at all. Some of them had led very-large-scale operations before, and would not have descended into Southeast Asia if not for the crackdown in China.

Feng himself has proved pivotal in holding all these people together. His ability to balance big-picture strategy as well as specific customer experience has earned him a reputation. Detractors accuse him of being too much of a micromanager, while people who have stayed find it intellectually stimulating and rewarding to constantly be challenged by him. This has helped Shopee dis-till a good core team as it has today.

Overcoming Growth Challenges

Of course, there are downsides. When things are going well, people praise you for your efficiency and effectiveness. But when things turn sour, the tone may switch quickly. Unfortunately, anyone who has followed large companies in recent years knows that we are in a much more nuanced world with a lot of undercurrents, from political, social, and geopolitical forces.

As a company operating in multiple continents and cultures, Shopee has managed to keep its internal culture and communications consistent and effi-cient; however, it can't be immune to some of these undercurrents. Society expects it to carry the social responsibility that is commensurate with its size and influence. While the group has been helping Shopee with some of these functions, as mentioned earlier in this chapter, it is probably not enough for it to be completely isolated from them.

While these are challenges that you would imagine Feng is not personally comfortable resolving, it is also important to note that he has an almost innate ability to identify the right priorities at the right time. It is highly possible that he tackles these issues only when the timing becomes right and these issues become actual issues—which might be now.

Ultimately, as the group leveraged the opportunity of cheap capital during the pandemic to catapult itself into a global operation, it is now at the stage of dealing with the challenges that come with it, and especially at the time of publishing this book, the bearish investor sentiment.

If these challenges are dealt with well, the company will end up having much more motivated senior ranks, and greater prospects as a global challenger in ecommerce and digital financial services.

Product and Beyond

Many consumers around us often complain by asking: Why would anyone use Shopee, as the product is still not the best in the market?

Truth be told, in a lot of finer details Shopee is still behind Lazada. At the time of this writing, Lazada's product search gives very accurate matching, while Shopee's often comes out with irrelevant items—for example, a search for bolts and nuts is supposed to only give you results in hardware, but in Shopee pistachios and almonds will appear.

Similarly, ShopeePay's experience, at the time of writing, is still not yet comparable to GrabPay's, run by Grab, Shopee's largest competitor in digital financial services in the region.

However, these probably did not matter so much in the past, as we mentioned earlier on Feng's strategy of only addressing the right problems at the right time. Shopee's primary focus has been young, urban, and mass consumers, while more sophisticated, highly demanding, and older customers are next in line to be tackled.

Would the hitherto target audience care so much about making search results 100% accurate versus getting the best deals for the goods they actually want to buy?

Another important aspect of Shopee's product is its large team of product managers. Compared to its main competitor Lazada, where many product managers boast years of experience working for Alibaba, Shopee's product team is still relatively raw.

One way to compensate for that is a combination of the sheer number of product managers and the close attention to product details of senior executives, including Chris Feng. One person familiar with the product management process of both companies told us that typically a Lazada product manager is probably five times as productive as Shopee colleagues, but with the numbers and streamlined command structure, Shopee makes up for this shortfall.

Gradually, as trusted lieutenants rise in the ranks to lead country markets, they are also empowered with more decision-making rights on product, marketing, and operations.

FINAL THOUGHTS ABOUT SHOPEE

Remember the quote from the São Paulo–based Chinese tech executive we mentioned at the beginning of this chapter? "Shopee manages its Brazilian workforce much more effectively than all the Chinese tech companies I have seen."

Now the reasons for his observations should be quite apparent. Shopee is in the best position to adapt some of the leading experiences on both ecommerce and digital financial services in various emerging markets. They learned from Chinese players from strategy and business models to operation. Meanwhile they kept its system open and flexible, to understand the local markets' development stage and real need. Over the years they have developed a unique compatibility, such as cross-culture management in different countries from Southeast Asia to South America, which enables them to be in a better position when competing with indigenous Chinese ecommerce players.

The question is whether Shopee, or Sea Group in general, will eventually turn profitable while continuing to grow their market size and share. At the time of writing there is a sense of skepticism from investors, some of whom argue that the market has already been adequately penetrated, with only limited future growth opportunities, while some others believe that like many emerging markets tech growth stories, profit will remain elusive for a long, long time.

We tend to agree with a third narrative—that with its current leadership, people, organization, and product, Shopee is in a great position to capture Southeast Asia's ecommerce prize. The company's execution, if Feng is still at the helm, should give investors confidence. If they fail, the problem is probably more fundamentally about the market than about competition or execution.

Global expansion offers a way to unlock a much larger addressable market, while hedging the risks that are inherently associated with emerging markets (political/policy uncertainty, currency exchange, and others). Whether Shopee can tackle global expansion effectively will provide an immensely useful case study for any company, tech or nontech, looking to create global domination.

J&T EXPRESS

Shopee is not alone in adapting Chinese ecommerce-related experiences across the globe. J&T Express, founded in Indonesia also in 2015, has become one of the largest courier companies in China, occupying a close to 20% market share in the highly competitive Chinese logistics market. At the time of this writing, J&T is making aggressive inroads globally, into markets including Brazil and Mexico, North America, Europe, and the Middle East.

J&T Express was started by Chinese businessman Jet Lee who, prior to founding the logistics company, spent years distributing smartphones for Oppo in Indonesia. He and his team had acquired invaluable experience and knowledge about the Indonesian market through working with distributors, resellers, and retailers in even the remotest islands of the vast archipelago.

The founding of an ecommerce logistics company was a natural extension—ecommerce was booming and no incumbent seemed to be doing a very good job. From its early days, J&T learned from the business models and operational systems of Chinese courier companies such as SF Express, even hiring some of the ex-SF cadres to help build its own systems, processes, and people management.

That, coupled with J&T's deep understanding of the market, its leadership's commitment in Indonesia, and the ability to manage a large local workforce, did wonders. Crucially, J&T also leveraged Shopee's rise, delivering parcels for the ecommerce platform at the scale and service level others simply could not match.

Many have asked us whether J&T and Shopee have a deeper relationship than just partnering on logistics. Our sense is that it is a pure commercial relationship, but it is a matter of convenience. Both have the Chinese gene, and yearn for rapid growth—working with each other to realize that growth is just natural. Shopee's recently fast foray into its own logistics service ShopeeExpress also suggests that the J&T partnership is now an in-depth marriage.

Very quickly, J&T became one of the largest courier companies in Indonesia. Through the Oppo network, it also quickly expanded to the rest of Southeast Asia. Its most audacious move, however, was launching an assault in the courier market in China, which by that time was dominated by SF Express and four companies invested by Alibaba. Competition was fierce and margins slim.

The timing was right. Because SF Express focused on the more premium segment, and while the other four companies all had investment from Alibaba, Pinduoduo—the fast rising challenger to Alibaba—needed a reliable partner to work with.

So the Shopee partnership in Southeast Asia was quickly replicated in China, with J&T Express growing volumes very fast in the vast market.

The other experience that was replicated was the reliance on Oppo's network of partners and distributors, allowing J&T to quickly expand its acceptance points.

Within a year after its launch, J&T was already delivering more than 20 million parcels every day in China.

Now, with Shopee investing heavily into Latin America, J&T also expanded into Brazil and Mexico. The other markets J&T has launched into include the Middle East and Europe—and the company is quickly building an ecosystem based on its logistics core.

 ## RESTEERING THE WHEEL FROM OUTSIDE CHINA

Both Sea Group and J&T Express exhibit the same POP-Leadership traits of major Chinese tech firms. Both have adopted Chinese business models, experiences, and leadership practices in emerging markets successfully.

A friend in a large emerging market who has worked with many Chinese tech firms shared: "These two companies, plus Lalamove, feel just different compared to the Chinese companies we have seen." The key difference, he says, "is that they seem to be able to manage their local organization and people really well, which Chinese companies often find hard."

This goes exactly to the points we discussed in the leadership, people, and organization chapters. Had these companies first emerged in China, they might not have been able to adapt to the realities of other markets so quickly.

In the case of J&T, if they had first started in China, they might not have even survived or convinced investors to invest money with them. Think back to the fate of community group buying startups mentioned in Chapter 8.

More companies like these will emerge in the future, as business models, ideas, and talents spill over across borders. We should not be surprised if these companies become a bigger force than proper Chinese tech companies in the global arena.

CHAPTER TEN

Connecting the Dots

O UR EARLIER CHAPTERS LARGELY take the perspective of Chinese tech firms, but in this chapter, we switch the perspective to local players to understand the potential externalities or spillover effects caused by investments from Chinese tech firms and their global presence.

On the one hand, local regulators and communities may be concerned about the potential "crowding out" effect on local indigenous firms due to the Chinese firms' stronger technology, greater resources, branding, and human capital. On the other hand, we should not forget that these foreign investments also create positive externalities through knowledge diffusion, technology transfer, and improvement of local infrastructure.

Currently, we see a large number of Chinese companies stepping out of their domestic market. While many are struggling in their global expansion, their entry has created new opportunities.

What are the implications of these Chinese players for local players? How can one replicate Chinese business models in their specific market? What is the validity of the Chinese players' operations and boundaries of applications? How can we connect the dots to draw lessons relevant to all of us as the readers of this book?

REPLICATING ALIPAY IN SOUTHEAST ASIA

Since 2016, Ant Group has invested in a number of e-wallets and mobile payment joint ventures across different countries in Southeast Asia. The strategy at that time was clear: Ant would provide the technological, product, and operational expertise needed for these companies to win at mobile payment services in their respective markets.

Ant had realized that it would be difficult for them to secure licenses and operational partnerships in those countries alone—any regulator would be very hesitant for a large foreign group known for aggressive tactics to have great control over the payment infrastructure and data of their country. And here we are talking about Southeast Asian regulators, who still vividly remembered the 1997 Asian Financial Crisis.

Therefore, Ant decided to work with strong local partners: CP Group in Thailand; CIMB Bank in Malaysia; Emtek Group in Indonesia; and Globe Telecom, an affiliate of Ayala Group in the Philippines (as we discuss in Chapter 7). They are either among the largest conglomerates in their respective countries or are in control of significant infrastructure that a mobile payment system can launch upon.

Another possible calculation of Ant is the following: If all these leading players adopt Ant's technology, Ant could build a powerful cross-border payment settlement network that puts itself akin to the role of Visa or Mastercard.

However, Ant and its partners are not the only giants, Chinese or local, that had similar plans for Southeast Asia. Ant's parent company—Alibaba's ecommerce competitor JD.com—quickly secured a partnership with Central Group, a main conglomerate rival of CP Group in Thailand.

The biggest competitor of Ant's Alipay in China is Tencent's WeChat Pay. Figure 10.1 shows that they are both super apps that provide more than just financial services after years of development. Tencent initially tried to launch its WeChat wallet in Malaysia (for which it had acquired a license) but also quickly realized that a joint venture or investment approach made more sense—for example, it backed Voyager, the digital finance venture of Smart, the main competitor of Globe in the Philippines.

Many other local conglomerates, tech companies, telcos, and media groups in Southeast Asia launched their own mobile payment wallets as well. At one point we counted more than 800 active wallets being promoted

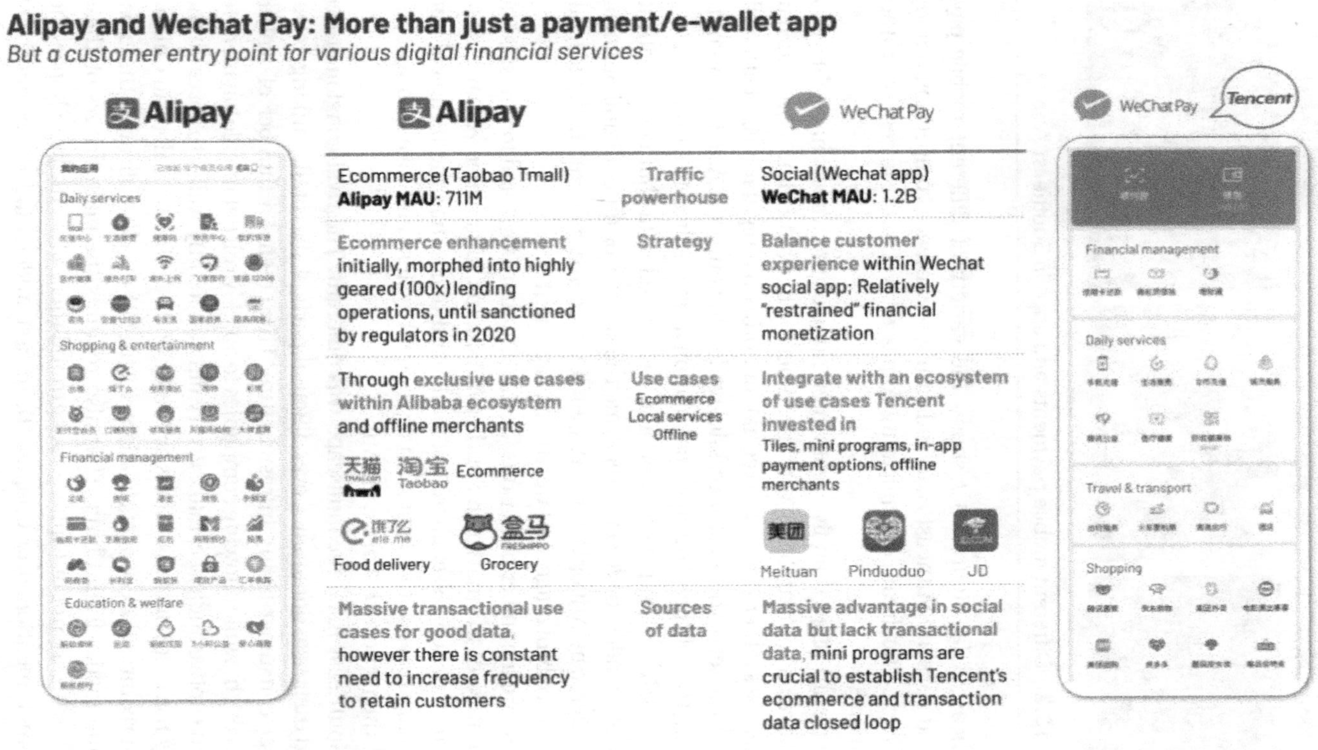

Alipay and Wechat Pay: More than just a payment/e-wallet app
But a customer entry point for various digital financial services

	Alipay	WeChat Pay
Traffic powerhouse	Ecommerce (Taobao Tmall) **Alipay MAU: 711M**	Social (Wechat app) **WeChat MAU: 1.2B**
Strategy	Ecommerce enhancement initially, morphed into highly geared (100x) lending operations, until sanctioned by regulators in 2020	Balance customer experience within Wechat social app: Relatively "restrained" financial monetization
Use cases / Ecommerce / Local services / Offline	Through exclusive use cases within Alibaba ecosystem and offline merchants	Integrate with an ecosystem of use cases Tencent invested in Tiles, mini programs, in-app payment options, offline merchants
Sources of data	Massive transactional use cases for good data, however there is constant need to increase frequency to retain customers	Massive advantage in social data but lack transactional data, mini programs are crucial to establish Tencent's ecommerce and transaction data closed loop

Alipay: 天猫 Tmall, 淘宝 Taobao — Ecommerce; 饿了么 ele.me — Food delivery; 盒马 FRESHIPPO — Grocery

WeChat Pay: Meituan, Pinduoduo, JD

FIGURE 10.1 Alipay and WeChat pay

Source: Company websites; Momentum Works research and insights. © Momentum Works.

FIGURE 10.2 Different mobile payments and apps in Southeast Asia

in Southeast Asia. Figure 10.2 shows the variety of different mobile payments and apps in Southeast Asia, indicating the intense competition among these players. The dream of replicating Alipay or WeChat's success is a rather common one.

The result? More than five years later, none of the wallets has achieved a level of success comparable to China's. Whereas in China almost every urban consumer uses either Alipay or WeChat, or both, on a regular basis, with cash nowhere to be seen, the most successful wallets in Southeast Asia hardly count more than 10% of their country's population as monthly active users.

With such limited penetration, and the limited data that comes with it, these mobile wallets have yet to replicate the success of other digital financial services offered by Ant or WeChat Pay, including consumer credit, digital lending, insurance brokerage, and wealth management.

There are a number of reasons for this situation. First and foremost is the fragmentation of the market in Southeast Asia. In China, when Alipay rose to prominence, nobody else, including banks and telcos, had the technical and operational expertise to create such a payment system and take it to success. Many of them chose to work with Ant.

In Southeast Asia, the landscape was different. The rival conglomerates each had their own dominance over different parts of the retail, banking, telco, and other consumer-facing infrastructure. This created a number of walled gardens which, while allowing companies to quickly roll out mobile payment to their own captive audience, make it very difficult for any wallet to cover all the everyday use cases or transactions as Alipay and WeChat have done in China.

That requires the e-wallets to adapt, instead of purely replicating the experience and tactics from China. That also requires the main drivers of these e-wallets, who are not of Chinese background themselves (unlike Shopee and J&T Express, mentioned in Chapter 9), to first understand the essence and inherent logic of the experiences in China and then determine how to adapt to be useful in their own setting.

When we asked the CEO of one of the Ant-backed mobile payment joint ventures in Southeast Asia about the real value that Ant has brought, we received a very interesting response:

> Contrary to popular belief, the superior technology brought by Ant's large product and engineering team was not that useful to us. The reason is that our reality is very different from what they have gone through in China—and it is very easy to miss that and develop features and products that are too complicated for our market.

He added:

> However, what we have found immensely useful is the expertise on strategy, business logic, and product sequencing. There were different paths we could take when we started this venture—to figure out what would be the best would involve detours, missteps, and crucial time wasted in a highly competitive environment. Ant's expertise has given us a clear idea of the pathways, possibilities, pre-requisites, and potential challenges or pitfalls—and we managed to roll out products and features in a successful sequence.

This is a very good summary of how to best make experiences from China useful by players in other markets, working with or competing against their Chinese counterparts.

 ## THE CASE OF SOCIAL COMMERCE

Since 2016, Southeast Asia has gone through a phase of replicating the successes of Chinese tech business models—fueled by investor belief that such models would succeed in a region that has significant geographical, cultural, and historical connections with China. E-wallet and mobile payments are just a couple of the many examples. Similar attempts have been made across ecommerce, logistics, local services, marketplaces, and other sectors.

Not only in Southeast Asia, but also globally, investors and entrepreneurs have been trying to take inspiration from Chinese business models. India, which now has a frosty relationship with Chinese apps (it has banned hundreds of Chinese apps, including TikTok and WeChat, since 2020), saw even more entrepreneurial zest in learning and adapting Chinese models (and seeking Chinese investment).

Terms that originate from China's hypercompetitive tech sector, such as O2O (offline to online), C2M (consumer to manufacturer), and social commerce, have become familiar to investors and entrepreneurs in different corners of the world.

Quite often, such terms are misunderstood or misused by observers and sometimes even the companies making use of them. Let's take the case of social commerce—where an array of companies, from Meesho in India to Facily in Brazil, have raced to replicate successes of Chinese companies such as Pinduoduo, Xiaohongshu ("little red book"), and even TikTok.

Perhaps the most intensive battlefield was Indonesia, where investors have poured money into a large array of companies all calling themselves social commerce. But what exactly is social commerce? How did it succeed in China? What is needed to succeed in Indonesia? Or would it succeed in Indonesia at all?

To understand this, we need to first understand what social commerce is. Ecommerce, like many other sectors in China, has evolved into a sophisticated environment where two ecosystems dominate. On the one hand is Alibaba, whose Taobao and Tmall platforms dominate the majority of formalized ecommerce; on the other hand is an ecosystem of major companies Tencent has invested in, including JD.com, Pinduoduo, and Meituan. TikTok, with its focus on video content, has also ventured into ecommerce, converting part of the billions of hours its users spend on the platform into buying time.

These are just the platforms—ecommerce in China also boasts a whole ecosystem of enablers, service providers, and most importantly, manufacturers and sellers. This book has thus far focused on the competition among major tech companies, but the competition among manufacturers, sellers, and brands is probably even more intense—for consumers' attention, and wallet share.

With platforms such as Alibaba's Taobao and Tmall, sellers compete by paying for advertising (through Alibaba's ad exchange, Alimama), offering discounts or low prices, and participating in shopping festivals and/or live streaming (which means more discounts and low prices). As a result, the customer acquisition cost and distribution costs for these sellers can be very high and margins thin—they are constantly seeking ways to break through and lower these costs.

Enter social commerce—an array of different methods to acquire and retain customers, and distribute goods online with cheaper costs, using social influence and social networks.

We have categorized social commerce into essentially four categories; to make these more relatable to the audience, we use some of the following global platforms:

- **Content:** Influencers (or sellers themselves) build content on popular platforms such as Instagram, Facebook, and TikTok, with consumers being able to place orders either directly or through a link to the ecommerce store; Xiaohongshu ("little red book") in China is a good example, where beauty influencers in particular write reviews and recommendations.
- **Reseller:** Members of a social or community network enlist to resell products to their network, often through social media and chat groups. Fashion and home products are often sold in this way—Yunji in China and Meesho in India are typical examples of this model.
- **Social group buying:** Buyers share with existing social network (through social media and chat groups as well) to form group deals at a discount; the rapidly ascending story of Pinduoduo, which we discussed in the leadership chapter, is a prime example.
- **Community group buying:** Buyers join a local community coordinated by a leader to make group buying purchases. In this model, the categories on sale are usually the most common denominators such as fresh produce and fast moving consumer goods (FMCG). We discussed the ongoing intensive competition of community group buying in China in Chapter 8.

While these models are different from each other, they all use social and community networks to reduce customer acquisition and distribution costs.

Now we project the business models in Indonesia, where an (increasing) number of players have adopted different strategies and models of social commerce. With a good understanding of the experience in China, the fundamental question of whether these businesses will succeed will come down to a few factors:

- We are sure that Indonesian consumers are deeply social—but do they have the willingness, tools, and consumption power to sustain large social commerce platforms? For example, WeChat in China makes social sharing, logistics, and payment possible through a single app. WhatsApp and Instagram, the social tools most commonly used by Indonesia, do not have payment or logistics capabilities to be able to close the transaction loop.

- What are the supply and distribution networks for the various categories of goods to be sold on social commerce? Take the example of FMCG; the traditional distribution networks are already very established. The very thin margin across the value chain makes it harder for social commerce players to derive any cost advantage, even if they can move large volumes. Similarly, for fresh produce, the gross margin is higher (up to 70%), but the supply is often unstable and wastage is a big problem. Traditional distribution channels rely on farmers, distributors, truckers, and final resellers absorbing costs and part of the losses to enhance the overall resilience of the network; when social commerce platforms try to have the distribution more organized, they have to bear with all these losses and inefficiencies. Even in China, with billions of dollars invested in community group buying, players are yet to make a profitable case. Players in more frictional markets such as Southeast Asia need to be extra operationally efficient.
- What would be the stance of major consumer tech platforms, such as Shopee, Grab, and Gojek? In China, areas where major tech platforms have entered generally have become very difficult for startups and new ventures to launch in, because major tech platforms often have economies of scale and lower cost capital, which are very difficult for startups to match. A quarter of widening losses for major tech companies can be compensated by growth in other areas or a public market raise, while for startups it is life and death.

When we talk about major tech players in Southeast Asia's context, we also need to be aware of players, including TikTok and Shopee, that are either Chinese or deeply inspired by the Chinese experience. They have not only more resources and firepower, but also the ability to internalize and adapt successful tactics and experience from China deeper and faster.

As this book was written, TikTok was making repeated attempts to tackle ecommerce in Indonesia, despite earlier challenges due to lack of an ecosystem including suppliers, trained influencer networks, and professional content creators. Shopee has been beefing up its own video capabilities to counter potential competition from TikTok, while at the same time offering its sellers a better way to showcase their products (in addition to text descriptions and photos).

The dynamics that are happening in Indonesia could have global ramifications and define how ecommerce will evolve globally (that is, outside the unique context of the Chinese market). Stakeholders should watch the evolution very closely.

MAKING ALL THIS RELEVANT

We have used two examples in this chapter to illustrate how players in other markets are able to make good use of experiences from China, and how Chinese-inspired competition in a particular sector (social commerce) and region (Indonesia) could have global ramifications.

These examples, when put in conjunction with the people, organization, product, and leadership (POP-Leadership) aspects we discussed in Part Two, as well as the experiences of Shopee, which represent more successful adaptation of Chinese models outside China by the global Chinese community, give us a complete picture of the influence of Chinese tech and internet firms in global markets.

A question we are often asked is: As investors, partners, competitors, employees, or even regulators of Chinese or Chinese-inspired tech globally, how do we make all these relevant to what we do?

Next we attempt to offer some basic pointers for different types of stakeholders.

Investors

In such a fast-moving world, tech and internet investing is a very tough profession. Investors need to not only constantly grasp the rapidly evolving dynamics, but also make decisions based on partial information and very short decision cycles.

The availability of various data-tracking services gives investors a good sense of the current dynamics in the market—from market share movements and consumer responses to marketing and promotional effectiveness. Momentum Works' Insights business unit also runs investors services, where we have provided investors the necessary on-the-ground insights for them to make more informed decisions. Through extensive discussions with various investors, from hedge funds to permanent capital, we have come to realize that while data provide a good level of comfort for analysis and decision, purely relying on and analyzing data alone can often give misleading results.

Despite the fact that we all are dealing with real-market data, there are still many more insights that outside investors might not have access to. Hard data will not tell whether the tech companies can act on the insights they have, provide the right leadership, mobilize the right people, evolve the right organizational structure, or constantly adapt their product to changing market conditions.

Investors should also focus on understanding the organizations they are looking at or investing in, the people leading them, and the organizational weaknesses and core competence to connect the dots and make informed bets.

That applies to venture capital investors who are looking at new startups in global markets, too. Experiences in China and history offer plenty of lessons on what the pitfalls might be, and what types of leadership are needed for success.

Corporate Leaders

Facing disruption, especially by players employing business models that have not yet existed in your market, how should you, as corporate leaders, respond effectively? Or more fundamentally, what are the options your company should pursue? Should you explore joint venture opportunities, expand your own businesses, or compete head on?

By now, you should realize that the considerations for corporate leaders here are essentially very similar to what Chinese business leaders face when they try to decide on their global expansion—the decision is just the opposite side of the same coin.

Unlike corporates in more established markets and sectors, tech companies in China typically shun traditional consulting firms, because they often do not believe consultants are sufficiently well informed about the fast-changing dynamics they are facing to be able to offer relevant or up-to-date best practices.

"We do not even do strategic planning, that's what corporates in mature industries or consultants do," a key division leader of a large tech company once told us. "What is the use of putting all these forecast numbers in a market where dynamics change on a daily basis?"

His alternate approach is top down. "We estimate the potential market size by a very rough set of assumptions, and then set a target for ourselves to hit. Once we launch, we just focus on solving problems along the way."

How do you work with, or compete against, people with such a mindset and aggression?

In fact, we were once asked by a large incumbent ecommerce player in a market where Shopee had just ventured in: How should we respond to Shopee?

"If we play by their tactics, we will lose a lot of money in a short period of time, with uncertainty about whether we can beat them at their game, or sustain ourselves to the day when they exhaust their resources and cease their attack," the questioner went on. "If we do not respond, as we are now, we see them chipping away our market share, day by day, without knowing when they will stop, or if they will ever stop."

As Chinese and Chinese-inspired tech companies are disrupting more and more sectors across the world, learning and adapting along the way, we are facing more and more such questions.

To find answers to these questions, it is essential to first gain a deep understanding about these Chinese companies, including their strategy, inherent business and operational logic, and their leadership. That understanding does not come from putting your own leaders in a strategy course or a high-pressure workshop, but by exposing them to real, in-depth examples, case studies from those tech companies. Most corporate leaders are exceptionally good at connecting the dots—they just need to get the right dots to connect.

One area that established corporates could definitely learn from these Chinese or Chinese-inspired tech companies is how to find the right people, inspire them, and keep the organization flexible enough to adapt to change—in Alibaba's lingo, "Change is the only constant" or "embrace change," as we discussed earlier. This is the corporate culture and organizational mindset that leaders need to consciously build, such that whenever they need to respond to something, they are not bogged down by organizational inertia.

A good example of how to respond to change involves the two leading banks in Thailand, Siam Commercial Bank (SCB) and Kasikorn Bank (K Bank). Like traditional lenders in many markets, they were comfortably profitable through a good offline presence, good brand names, and a large asset base.

Ant Group's investment in CP Group's AscendMoney served as a big wake-up call for the two banks. While their counterparts in many markets battled with internal organizational inertia and slow decision making in shifting strategies to respond to disruption, the two showed exceptional leadership in first understanding what disruption was coming their way, and then responding actively. They strengthened the leadership team, restructured the organization, actively absorbed the top talent, and experimented with a slew of different digital financial services products from mobile payment to digital lending.

Not only in the core domain of their finance businesses, through investment, joint venture, and their own corporate venture building, they have actively expanded their sphere into food delivery, data analytics, and even crypto platforms. As of the end of March 2022, SCB's venture arm SCB10x boasted an investment portfolio of 51 companies and 5 subsidiaries across 13 countries. Through seeding many leading venture capital firms focused not only on Southeast Asia but also on China, SCB has been able to learn about the upcoming business models and devise corresponding strategies quickly.

Entrepreneurs

Entrepreneurs in almost every tech sector can find reference points in China, where the leading players have fought a long and hard war against competition as well as adjacent players.

There are a tremendous number of lessons to be learned about how a particular business model could evolve, the different pathways with associated opportunities and challenges, and operational considerations to optimize each part of the business to stay more competitive.

In addition, entrepreneurs would also benefit from understanding the personal journey, as well as thinking and evolution in leadership, people, organization, and product from their more experienced Chinese counterparts, described throughout this book.

Many of the lessons drawn from history and revolutionary leaders would be useful to entrepreneurs who often have to make very tough decisions with imperfect information to take their business forward, and deal with the emotional and psychological challenges along the way.

Regulators

The regulatory actions toward Ant Group in 2020 seemed to be long overdue, as various regulators had voiced concerns about areas of Ant's business, notably the excessive leverage for its consumer lending products, for a long time.

Notably, in areas such as P2P lending, insurtech, and other financial technologies, the regulators initially did not institute a clear stance or licensing scheme, but came in to crack down later on when the sector boomed.

Regulators in China were often in a tough spot. When many of the tech-related businesses were growing fast, regulators did not really have a global reference point to study and set rules accordingly. The fact that many such businesses grew very quickly posed extra challenges to regulators—if they set the rules too tight at the onset, they stifle innovation; if they do not, things might quickly evolve out of control and drastic actions are needed to rein things in. It is a tough trade-off for many regulators on whether, when, and how to regulate the new tech business.

In China, more concerted efforts came in in 2021. Antitrust enforcement saw Alibaba, Meituan, and Tencent fined for previous offenses, including forcing merchants on the platform to not work with rival platforms, or merging subsidiaries that create monopolistic market positions; Ant's various businesses were investigated and decomposed to put under more stringent regulatory buckets, with the group transitioning to a financial holding company subject to similar regulations on banks; data security and privacy of various platforms were seriously scrutinized, with regulatory actions applied accordingly; a new scheme to classify individuals who leverage digital platforms to make a living (gig workers, ecommerce sellers, etc.) into different groups and avail them of social security and protection accordingly.

While we think most of these actions are overdue, regulators outside China can learn from the experience in China in two ways:

1. Knowing how a tech sector, if left unchecked or lightly regulated, might evolve—allowing regulators to apply the right levels of checks and balances early on.
2. Understanding the rationale behind recent regulatory storms in China, comparing the differences in circumstances between China and their home country, and deciding the best regulatory courses that suit the country's specifics.

With all these rich examples from China, regulators elsewhere can learn from the mistakes, better preempt or keep things in check, and ultimately better perform their balancing act to foster the healthy growth of tech-enabled innovation.

The Rest of Us

Many executives we know working for international firms such as Google, Facebook, and McKinsey have been approached by Chinese or Chinese-inspired tech companies for potential job opportunities. Young talent often sees their friends and schoolmates joining these tech companies.

While the likes of TikTok have adapted their culture to be more international, most major Chinese tech firms still boast their particular culture that defined their original success in China. Foreign employees, even ethnic Chinese, often find cultural shock upon joining these companies, which we discussed extensively in Chapter 5, People.

However, we all know that to develop and continuously grow a career, comfortable places are dangerous. Is the discomfort or cultural shock in Chinese firms something that one should get used to (or at least get some exposure to) for the greater benefits of long-term career development and personal wealth accumulation?

There is no better way to understand these companies than by observing and judging their leadership, people, and organization. That will give you the ability to make an informed decision on whether to join a Chinese or Chinese-inspired tech company when presented with such an opportunity. This will also give you the ability to evaluate such companies as partners, customers, or suppliers.

Ultimately, life is an investment. Through a series of the right decisions, you can compound the returns and move much ahead of your peers. You certainly do not want the reverse to happen.

 ## SUMMARY

What we have discussed in this chapter are some of the most obvious but also most important aspects for understanding the POP-Leadership of Chinese tech companies that will benefit different profiles of readers in the tech ecosystem.

As the ecosystem is still constantly and rapidly evolving, it is important to evolve and adapt your understanding as well. We will keep monitoring the tech ecosystem, and we welcome any discussions with the audience on a continuous basis. We will continue to share some of these discussions through various channels, including the blog: www.pop-l.com.

CHAPTER ELEVEN

Cross-Pollination of Global Markets

ALTHOUGH OUR BOOK STARTS with a focus on Chinese tech firms in their overseas expansion, they are not alone in their quest to seize opportunities globally. Truth be told, the fact that some of the major Chinese tech companies such as Alibaba and Tencent could grow to their level of success today can be attributed to earlier pioneers who started seeking global tech opportunities more than two decades ago.

South Africa's Naspers, Japan's SoftBank, and Germany's Rocket Internet were among some of the key players of that era. All of them are still active to this day. Now we are seeing much more exchange of business models, product and operational expertise, and capital and talent across all the major markets globally.

A TRIPLET OF PIONEERS IN SEEKING GLOBAL TECH OPPORTUNITIES

Several Chinese tech firms accomplished the miracle of quick growth to join the "100-billion club" in company valuation. Undoubtedly, the leadership, people, organization, and product parameters all played a critical role in the firms' historical development. Their journey was not smooth, however, especially in the early stages. Without funding, knowledge, and human capital from earlier pioneers in searching for global opportunities, it is hard to imagine that Chinese tech firms could stand as they are today.

How Naspers Discovered Tencent

As the year 2000 was close to its end, Pony Ma's new product, OICQ (later renamed QQ), was rapidly gaining popularity, as well as server costs. The team was close to depleting the US$2.2 million angel investment, but nobody across China was willing to put in more investment because "nobody in this world knew how it can make money," according to someone familiar with that part of the company's history.

When Pony Ma became increasingly desperate, David Wallerstein, an American executive representing a South African firm that Pony Ma and his team had never heard of, showed up.

Wallerstein, who spoke fluent Mandarin, was working for MIH, an investment subsidiary of South African media conglomerate Naspers. He discovered Tencent by chance, as he saw young people across China's internet cafes put OICQ on their desktop, and many founders of businesses pitching him for investment printed OICQ numbers on their business cards.

MIH quickly struck a deal to invest in Tencent, acquiring 46.5% of the shares to become the largest shareholder. Naspers, through its Amsterdam-listed subsidiary Prosus, remains Tencent's largest shareholder to this day.

In April 2021, when Naspers liquidated a portion of its shareholding, the realized return was roughly 7,800 times, making it one of the most successful investment bets in history.

In fact, Tencent, while the largest and arguably the most successful, was far from the only bet Naspers made in emerging markets. Since the mid-1990s, the group has been systematically hunting for tried and workable business models across different markets.

In 2010, Naspers acquired OLX, an online classified site (an alternative Craigslist) that was operating in more than 30 countries across the Americas, Asia, the Middle East, and Africa; in 2010, it started a payment company in Africa later renamed PayU, which consolidated payment assets to expand into 17 countries.

Even in China, Naspers had invested or launched a number of bets early on in addition to Tencent, although none of the others was particularly successful.

In recent years it also moved to invest into some of the more operations-heavy business models such as food delivery, where Naspers has acquired stakes in India's Swiggy, Latin America's iFood, and Frankfurt-listed Delivery-Hero group.

A similar bet to that of Tencent was the acquisition of a 30% share of Mail.ru (later renamed VK Group), one of the top internet groups in Russia.

SoftBank, Alibaba, and the Time Machine

Masayoshi Son, the Japanese billionaire tech visionary, is now a household name, after his ambitious $100 billion Vision Fund made aggressive investments into companies including WeWork and Uber.

There were stories of SoftBank executives going to founders, offering so much money they could not resist, and threatening to give the money to competitors if they hesitated.

Vision Fund also bets on similar companies across multiple continents. In addition to Uber, it also invested large sums into Didi in China, Southeast Asia's Grab, and Ola in India. In used car marketplaces, it has invested in India's Car24, Southeast Asia's Carro, and China's Guazi. In food and other local on-demand deliveries, the Softbank Vision Fund counts DoorDash, China's ele.me, and Latin America's Rappi in its portfolio, in addition to many of the ride-hailing companies it had invested in that also went into deliveries.

SoftBank, which made its name globally through an early investment in the Alibaba Group, has been well known in China for years. This investment made SoftBank Alibaba's largest shareholder and is comparable to MIH's investment in Tencent in timing and magnitude of success.

However, there is also one Masayoshi Son anecdote that is well known in China's tech circle but little mentioned outside China—the "time machine" theory (as we briefly mentioned in Chapter 3). The idea is taking proven business models from mature markets like that in the United States to replicate in markets where tech is less developed.

Investment analysts, journalists, and even founders in China often mention this term to describe Chinese technology companies expanding into markets where tech is less developed.

However, outside China this theory is little known. So where does this theory come from? Is it real or another urban legend?

In SoftBank's annual report of 2000, there is this description in the "Strategy" section:

> SOFTBANK is pursuing a so-called "time machine management" strategy
> to foster the global incubation of superior business models found through
> its venture capital operations in the United States. This strategy has given

rise to numerous cutting-edge success stories in the Internet arena. A prime example is the world-renowned portal site Yahoo! SOFTBANK provided early stage financing for Yahoo! in the United States. It then extended the business model to Japan, South Korea, Germany, France, and the United Kingdom in partnership with Yahoo! Inc. in the United States.

SoftBank went on to describe how it intended to implement the time machine management strategy in Latin America, Europe, and Asia, as well as emerging markets in general, through a series of internet investments and joint ventures. In 2020, SoftBank had a portfolio of more than 400 companies across the world, including the ventures it had funded.

Which means that the idea of betting on global opportunities based on proven models and a perceived time window was entrenched in Masayoshi Son's mind, almost two decades before the launch of the US$100 billion Vision Fund.

While Masayoshi Son has moved on with his metaphors—he now describes SoftBank as a goose laying golden eggs (successful tech companies), much more lively compared to the time machine strategy—the spirit lives on.

Rocket Internet's Quest to Build the World's Largest Internet Platform

Masayoshi Son was not alone. In 2014, Rocket Internet, a Berlin, Germany–based group then famously known for ruthlessly copying successful business models, went public.

Its founders, the Samwer brothers led by middle brother Oliver, had already sold a few successful copycats by the time they founded the group in 2007, including auction site Alando to eBay in 1999 for US$43 million, just a few months after founding.

Rocket Internet went on to found companies including Zalando, inspired by Zappos, and CityDeal, which it successfully sold to Groupon. By the time of its own IPO, Rocket had a full suite of companies under its belt—covering ecommerce, marketplaces, ride hailing, and digital payment. It had presence through portfolio companies or joint ventures in Europe, the Middle East, Africa, Asia (even including China), Latin America, and even Iran.

The mission on its website at that time was "to become the world's largest internet platform outside the United States and China."

The company was as ambitious as it was controversial—many entrepreneurs around the world accused it of abusing its financial superiority to crush local entrepreneurial originality. Its response was that ideas are useless, it is the execution that counts, and Rocket just executes better than many others.

While typical venture capital firms shunned working with Rocket Internet, conglomerates, European family offices, and especially telecom companies around the world, which wanted to own a piece of the internet but did not have the capabilities to execute, ploughed money into the Group.

Many of its ventures were shut down very quickly, a decision indicating that the group was more interested in managing an adjustable portfolio of ventures than in becoming emotionally attached to any ideas (as entrepreneurs would do). During Rocket Internet's few years as a public company (it went private again in 2020), analysts constantly complained that they found it impossible to value the company because of its complex structure, especially the large number of shell entities at the mezzanine level between the group and its operating ventures.

Despite all these controversies, Rocket Internet as a group was indeed successful. After Zalando, it generated a number of listed companies, including Global Fashion Group (a collection of fashion ecommerce companies operating in Asia, Australia/New Zealand, Latin America, and Eastern Europe/Central Asia), Jumia (an Africa-focused internet group), DeliveryHero (food delivery), and HelloFresh (a Germany-based meal kit platform). It also sold its ecommerce platform Lazada and Daraz (focused on South Asia [i.e., India]) to Alibaba.

It also accelerated the development of infrastructure, talent, and ecosystem as a whole in many markets, paving the way for later entrepreneurs and investors. Many founders, including Shopee's Chris Feng, Gojek's Nadiem Makarim, and one of the authors of this book, had stints at Rocket Internet and learned greatly through the experience.

In fact, the Rocket Internet experience is always a cause of fascination to Chinese tech investors and entrepreneurs. As mentioned earlier, they often view copying as a virtue, and they prize swift execution.

In 2018, we spoke with the cofounder of a large technology company turned VC investor, who had just relocated to Singapore. After hearing the full story of Rocket Internet, he said: "What Rocket Internet had done, could and should have been done by the Chinese companies."

He paused, and then added: "If only we were not bogged down with the intensive domestic competition in China, and spoke better English."

CHINESE TECH FIRMS' GLOBAL MAP

Fast-forward to the late 2010s and early 2020s: now the Chinese tech companies are mature and big enough to start implementing their global ambitions.

While there are expansion activities across the world, we chose Southeast Asia, India, the Middle East, Africa, and Latin America to discuss various opportunities and potential challenges. This map would give you a good picture when you consider the question of where to start the journey if you have a similar dream of expansion.

Southeast Asia—the First Port of Call

Southeast Asia, with its combined population of 655 million people, is often the first port of call for Chinese tech companies' global ventures. This is often not a conscious decision based on Southeast Asia's being more attractive or more ready compared to other regions, but the fact that Southeast Asia is familiar to China.

Most Southeast Asian capitals are either in the same time zone or one hour apart with China, and direct flight time from Beijing to the farthest Southeast Asian capital is only six hours. This makes team coordination and movement of personnel much easier.

In addition, most Southeast Asian countries boast a large ethnic Chinese community, which often exercises large influence in business. A majority of the roughly 50 million overseas Chinese live in Southeast Asia, with Chinese family groups among the largest conglomerates in Singapore, Kuala Lumpur, Bangkok, Jakarta, and Manila. While Vietnam is a notable exception in ethnic Chinese influence in business, as a whole the culture and values are very similar to those of China, due to millennia of (often uneasy) coexistence. Figure 11.1 describes the six sizeable economies in Southeast Asia. Figure 11.2 presents their corresponding very healthy demographics for future consumption growth.

The region's business hub, Singapore, has an ethnic Chinese majority (75%), where Mandarin Chinese is one of its official languages and probably the one most spoken on the streets.

The geographical, cultural, and historical proximity makes Southeast Asia an attractive first market for Chinese tech companies. Alibaba, Tencent, ByteDance, and many other Chinese tech companies have set up their regional or global headquarters in Singapore; from Akulaku to Flash Express, the Chinese community has built many tech unicorns in the region; and almost all of the region's tech majors, from Sea to Grab to Gojek, count large Chinese tech groups as their key backers.

Southeast Asia is composed of a set of (six) sizeable economies
Each is big enough to be attractive, but not big enough to breed consumer internet giants in US/CN scale

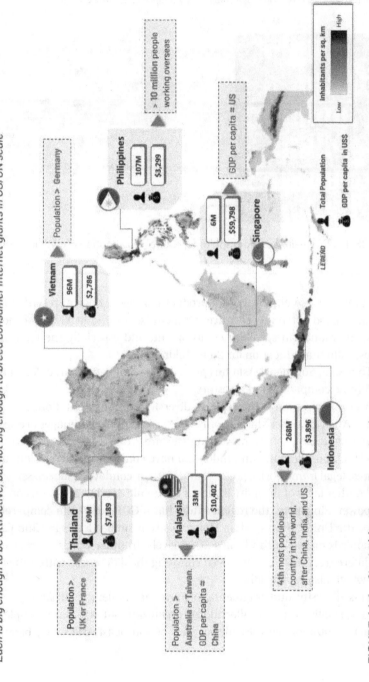

FIGURE 11.1 Six sizable economics in Southeast Asia

Source: World Bank; Momentum Works insights. © Momentum Works.

Southeast Asia has very healthy demographics for future consumption growth

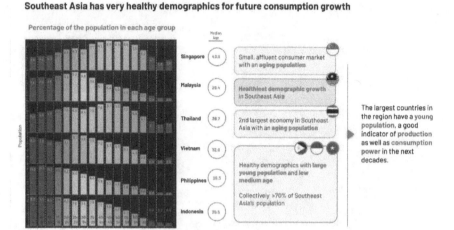

FIGURE 11.2   Healthy demographics for future consumption growth in Southeast Asia

Many Chinese engineers and product managers have also settled in the region. A good personal incentive for them is to settle in Singapore for the safety, the Mandarin-speaking environment, and especially the English and Chinese bilingual education for their children.

That said, Southeast Asia has proved to be a difficult market for many of the Chinese companies and investors.

A primary reason is the region's diversity and fragmentation. Compared to Latin America, where largely only two (very similar) languages are spoken, or the Middle East, where most countries are Arabic-speaking, in Southeast Asia, Thailand, Vietnam, and Indonesia have completely different languages, cultures, legal frameworks, policies, and ways of conducting business.

Another layer of diversity is that of economic development and consumption power. Singapore, the region's hub, has a GDP per capita comparable to that of the United States, while Malaysia's is at similar levels as China's, and Vietnam's is on par with China's 2007 levels. In other words, a spectrum of 30x difference, and that is not even counting the differences within the countries, which also vary wildly.

This diversity and fragmentation present challenges in strategic planning (especially resource allocation), organizational structure, people, and product localization. Historically very few of Southeast Asia's megabusinesses managed to dominate the entire region, as they had not much clear advantage compared to country champion conglomerates.

Another key factor: the Chinese urban middle class live in a cluster of large cities with close transportation links to each other: according to the latest census, conducted in 2020, there are 21 cities in China with a population of more than 5 million; 7 of those cities have more than 10 million residents. This concentration of population and consumption power, together with the largely similar urban planning, unified language, and the just-mentioned transport infrastructure, made winning consumer tech companies national-level behemoths.

In comparison, each country in Southeast Asia has its economic might concentrated in only one (or, in the case of Vietnam, two) megacities and their respective surrounding urban areas. While most of these cities are within a three-hour flight from each other, the challenging geography in the region, with mountains and archipelagos, makes effective transport and logistics links between these big cities difficult. Figure 11.3 shows the major cities in Southeast Asia.

The Chinese tech companies, which are used to pooling resources together to fight large, nationwide wars, now have to face a different set of contests, each of them much smaller in scale and requiring a lot of customization in people, product, and even strategy.

The fact that most of the region's large conglomerates are run by ethnic Chinese families and still controlled by the second or third generation instead of professional managers adds another layer of complexity.

On the one hand, many of these family groups have been investing in China for decades, and have a very good understanding of the business dynamics there. As a result, they value the experiences of, and are very keen to work with, Chinese tech firms.

On the other hand, precisely because of this understanding, these groups, as well as the other associated interests, are very careful about guarding their territory. Using the words of a friend from a large Chinese tech firm who used to run a joint venture with a Southeast Asian family conglomerate: "After two years of working together, I finally understood the real objective of our partner was to learn the skills from us to be future proof, rather than proactively disrupting their existing offline business."

Despite all these challenges, Southeast Asia, with the geographical, cultural, and historical proximity we mentioned earlier, remains attractive to

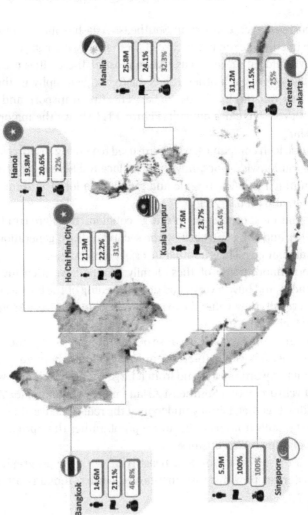

Megacities play a key role in developing tech in Southeast Asia
You can only tackle the vast remainder of the markets once you have secured the megacities

The 7 megacities represent ⅓ of the Southeast Asia's market in terms of business activities and consumption.

A megacity is a symbol of growth engine. Players that can't hold on to a megacity will lose out.

- Population of metropolitan area
- % of entire population
- % GDP in metropolitan area

Manila
- 25.8M
- 24.1%
- 32.3%

Greater Jakarta
- 31.2M
- 11.5%
- 25%

Hanoi
- 19.8M
- 20.6%
- 22%

Ho Chi Minh City
- 21.3M
- 22.2%
- 31%

Kuala Lumpur
- 7.6M
- 23.7%
- 16.4%

Bangkok
- 14.6M
- 21.1%
- 46.8%

Singapore
- 5.9M
- 100%
- 100%

FIGURE 11.3 Major cities in Southeast Asia

Note: Based on largest city by population in each country.

Sources: National census; Populationstats; Malaysia Department of Statistics; Philippines Statistics Authority; Thailand NSO; Philippines Statistics Authority; GRDP of Vietnamese provinces. © Momentum Works.

Chinese tech. Southeast Asia is the region where Chinese tech leaders and executives are willing to stay put for the long term to overcome all the challenges just mentioned.

The region also has seen the most success of Chinese techs' global ventures. Besides, both Sea Group and J&T Express, discussed in detail in Chapter 9, originated from Southeast Asia. Tencent is a key shareholder of both companies.

India—Involution and Geopolitics

Once upon a time, India was the second most attractive region for Chinese tech, just behind Southeast Asia. Its population was comparable to that of China, and would eventually overtake it in the early to mid-2020s. This metric alone makes India an important market that global tech giants and investors do not want to miss.

Alibaba indeed had ambitions in India. Alongside its shareholder Soft-Bank, Alibaba invested in Snapdeal, then India's second-largest ecommerce platform, as early as 2015. Alibaba also invested in Paytm, India's largest mobile and digital payment service, along with its fintech associate Ant Group. When Snapdeal lost momentum in the competition with Flipkart and Amazon, Alibaba even fostered an ecommerce joint venture Paytm Mall, which eventually did not achieve meaningful market share, either.

Through acquisition of Guangzhou-based UCWeb, at one point India's largest mobile browser, Alibaba also went into the content business in India, a traditional stronghold of its archrival Tencent.

Tencent, on the other hand, gradually brought over India portal ibobo .com from its parent Naspers from 2008 onward. After the failure to promote WeChat in the country, Tencent went on with more investments, into Flip-Kart, health platform Practo, travel portal MakeMyTrip, food delivery platform Swiggy, B2B ecommerce platform Udaan, chat app Hike, ride-hailing leader Ola, and social platform ShareChat, among others. In published deals, Tencent had put more than US$2 billion in the country.

Other Chinese companies went in as well, leveraging China's supply chain advantages, especially in electronics (in the cases of Xiaomi, OPPO, and One-Plus mentioned earlier), and rich experience in operating content and social apps. Among these, TikTok was particularly successful in amassing more than 200 million active users by 2020. NewsDog, a news aggregator run by a Chinese team but focused on India, received a US$50 million investment led by Tencent in early 2018.

A challenge, however, has been looming over Chinese investors' sentiment about the country. Despite the ease in acquiring a large user base, monetizing these users in India has been incredibly difficult. This is due to the low consumption power (GDP per capita of India was about one fifth of that of China in 2021) and underdevelopment of the online advertising market.

While everyone believed that in the long term, compounding economic growth would eventually bring this market to viable levels, few Chinese decision makers and executives had the patience to stay put in the market for a decade or even longer.

Another challenge is the intense domestic competition among Indian entrepreneurs. A good friend of ours, who represented a large Shanghai-based investor and made a number of investments in India, never put money with a single Chinese entrepreneur or team targeting the Indian market. His gut told him then that it was not a good idea, but he was only able to explain why in 2021, a few years down the road: "Tech entrepreneurship in India is a typical case of involution, leaving little chance for outsiders to outcompete the smart, capable, and aggressive locals." By then, "involution" was already a catchphrase in China, as described in Chapter 8.

Of course, the market sentiment was further hit by geopolitics in 2020 that took everyone by surprise. A deadly clash broke out between Indian and Chinese border forces in the remote, high-altitude Galwan Valley, resulting in multiple deaths on both sides. The next month, the Indian government announced the ban of nearly 60 Chinese apps, including TikTok, WeChat, and Bigo Live, but also South Korea's game *PUBG Mobile*, then distributed by Tencent.

While some public policy experts expected this ban to be temporary, India went on to issue two further bans, blocking in total more than 200 apps that were either Chinese or perceived to be Chinese-linked. In a move in February 2022, Sea Group's major game *Free Fire* was also on the list, due to perceived links with China. The Government of Singapore, where Sea Group is based, has sought clarification from the Indian government through official channels; however, the issue remains unresolved as of this writing.

This casts doubt on Shopee, Sea Group's ecommerce operations, which has been expanding in India. Moreover, even if this ban is eventually lifted, the confidence of Chinese tech companies and investors in the Indian market will probably take years to recover, if it ever does.

Middle East—Gaps Left by the Giants

When JollyChic, the Chinese cross-border ecommerce company we introduced in Chapter 3, first ramped up their expansion into Saudi Arabia in 2016, no other Chinese ecommerce player had paid serious attention to the region. To most of them, the region remained a mystery, even though the United States and other Western markets had become highly competitive.

However, Chinese wholesalers and retailers had been operating out of the Middle East for a long time. Dragon City, a large integrated development boasting more than 5,000 shops dedicated to Chinese manufacturing and wholesale businesses, opened its doors as early as in 2004. Estimates of Chinese businessmen stationed in the Emirate ranged from 100,000 to 300,000.

The region, which included the affluent Gulf countries but also large populations such as Egypt and Iran, has yet to attract significant attention from Chinese tech majors, primarily due to distance and the fact that in the Gulf, the most affluent part of the region, only half of the 54 million inhabitants are locals.

Small, entrepreneurial firms fill in the gap. Aside from JollyChic, a number of other cross-border ecommerce companies joined the competition in 2018; Rita Huang, an ex-Huawei executive, built iMile, a logistics company, to serve the region's booming ecommerce volume; content and gaming companies, including Yalla Group, Mena Mobile, and Singapore-based, Chinese-run Mozat have occupied their own lucrative niches.

In addition, many Chinese AI companies have also accepted invitations from governments across the Gulf to set up R&D centers and build dedicated products for the region. eWTP, an Alibaba-affiliated investor helmed by UCWeb founder Yu Yongfu, set up a joint venture with Saudi Arabia's sovereign fund PIF to systematically build and incubate tech joint ventures with other Chinese tech leaders.

In the current geopolitical environment where rich Gulf nations seek to diversify their reliance on the United States and build up their own domestic tech capabilities, expect more such collaboration to take shape.

Africa—for the Most Determined

In Chapter 7, we described how Transsion, a Chinese smartphone brand turned ecosystem player, localized its suite of products for the African market. Transsion has worked with a number of Chinese tech companies, such as gaming and content giant NetEase, to incubate joint ventures based on its large customer base.

Similar to the Middle East, Chinese businessmen have long been active in Africa—for example, after the civil war in Angola, the local government of the ancestral county of one of the authors of this book estimated that there were 3,000–5,000 individuals from the country actively in Angola, working in trade, retail, infrastructure, and hospitality. Chinese state-owned enterprises also invested heavily in the region's infrastructure and natural resources. A good friend of ours has been working for Huawei in Africa for more than 15 years, moving from Kinshasa to Lusanka, to Nairobi, and now Johannesburg.

Tech entrepreneurs are, however, late to the game. While many see the opportunities in the vast region (three times the land mass of China, with comparable population), few made the move.

The reason is actually simple—they had better choices in life and work. The infrastructure, food, and lifestyle in Jakarta, Indonesia, for example, are a few notches higher than in Lagos, Nigeria.

A few entrepreneurs we know who ventured to set up tech businesses in West Africa had big challenges recruiting experienced operation executives from China to join their team, without whom these businesses would not be more competitive than local counterparts. This is the people issue we discussed earlier.

This might have changed when Zhou Yahui, founder of Kunlun Wanwei, a very successful gaming group and a known investor, decided to venture into Africa himself. He had acquired the web browser Opera, which was popular in Africa, and decided to launch OPay, a payment, financial services, ride-hailing, and local service group in Nigeria.

Because of Zhou's personal appeal, both experienced Chinese tech talent and investors joined him. Many top-tier VC firms in China, from IDG Capital to Sequoia China to Gaorong and SourceCode, are shareholders of Opay.

Although Opay encountered a lot more challenges than Zhou had anticipated along the way, the billionaire entrepreneur is determined. He launched a public call to fellow Chinese entrepreneurs to join him in exploring African markets.

Zhou said that in places like India and Indonesia, a lot of US-educated founders returned with experience, technology, and access to capital, in addition to local resources which they already had: "Chinese entrepreneurs in these markets have no advantage at all."

However, while he was bullish about the long-term prospects of Chinese tech entrepreneurship in Africa, he cautioned that precisely because of the earlier stage of tech Africa is in, it would take years of patience and determination to build successful businesses there.

"You can't be opportunistic in Africa," he said. "You have to work through years, or even decades, of hard work to reap the benefits."

Latin America—the New Frontier

Latin America is the last of the regions in this chapter, but it has many similarities to Southeast Asia, the first region discussed earlier on.

Both regions have six major economies with comparable total populations and economic size. The largest economies in both regions, Brazil and Indonesia, have abundant natural resources, and draw in half of the venture capital investment in their respective regions. There are also other comparisons to draw between the two regions, shown in Figure 11.4.

Historically, Southeast Asia has received much more tech investment than Latin America (shown in Figure 11.5), especially in 2017–2018, when a number of megadeals (with an investment size of US$100 million and greater) shored up the region's tech majors. Companies in similar sectors often had to

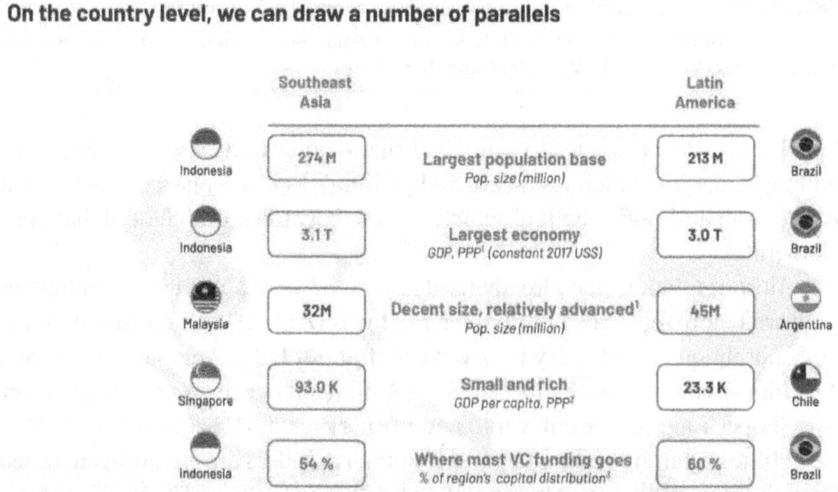

On the country level, we can draw a number of parallels

	Southeast Asia		Latin America	
Indonesia	274 M	Largest population base Pop. size (million)	213 M	Brazil
Indonesia	3.1 T	Largest economy GDP, PPP[1] (constant 2017 US$)	3.0 T	Brazil
Malaysia	32M	Decent size, relatively advanced[1] Pop. size (million)	45M	Argentina
Singapore	93.0 K	Small and rich GDP per capita, PPP[2]	23.3 K	Chile
Indonesia	54 %	Where most VC funding goes % of region's capital distribution[3]	60 %	Brazil

FIGURE 11.4   Comparison between Southeast Asia and Latin America
Note: 2020, unless specifically stated otherwise.

[1] Sufficiently deep markets for initial sustainability early on while venture capital is still scarce: hub for early regional players in Web 1.0: in Southeast Asia: MOL, Jobstreet. Out of Argentina: MercadoLibre.
[2] Constant 2017 US$.
[3] HT 2021, ex-regional companies.

Source: © Momentum Works.

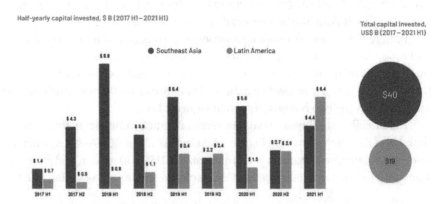

FIGURE 11.5 Investment in Southeast Asia versus Latin America

For more information, refer to Cento's Southeast Asia Tech Investment 2021 HT Report. *Source:* Cento Ventures; LAVCA. © Momentum Works.

look for an IPO or acquisition much earlier—one example was Brazil's ride-hailing leader 99, which was acquired by China's Didi, as opposed to Southeast Asia's Grab and Gojek, both of which grew to become multibillion-dollar conglomerates.

Although Baidu had already tried and failed with a number of initiatives in Brazil before (described in Chapter 7), it was Didi's 2018 acquisition of 99, and subsequent direct entry into Mexico, that put Latin America on the map for Chinese companies and investors. Another confidence, or urgency, booster was Shopee's aggressive entry into Latin America, which began in late 2019.

Chinese gaming companies and ecommerce sellers are encouraged by Sea Group's foray to follow suit in expanding into Latin America. TikTok, Kuaishou, and J&T Express also built a meaningful presence in Brazil, which is used as a base for wider Latin American expansion.

In the meantime, we are also seeing a number of Chinese-inspired business models, such as Stori, a fintech company in Mexico, and Facily, a social commerce platform in Brazil, expanding fast. GGV, a global investor that had made dozens of successful bets in China in the 2010s, is now working hard to replicate that success through Latin America's budding entrepreneurs.

The physical distance between China and Latin America (up to 40 hours' flight and transit time) seems to be gradually ignored as tech business connections between both regions evolve.

THE CROSS-POLLINATION OF GLOBAL EMERGING MARKETS

We have touched upon Southeast Asia, the Middle East, Africa, and Latin America to present a taste of Chinese tech and their influences across the world. There is much more happening.

In ecommerce, content, and social domains, many Chinese companies and entrepreneurs are targeting US and other mature markets, hoping to at least partially replicate the successes of SheIn and TikTok. Chinese cross-border ecommerce platforms AliExpress and Banggood count Russia as their biggest market prior to the war in Ukraine this year. Quite a number of Chinese fintech and content entrepreneurs are building their businesses in Turkey too.

Again, it is perhaps not surprising that the Chinese are not alone in this effort to expand globally. Indian and Russian entrepreneurs and capital have built meaningful businesses in Southeast Asia, while founders of Latin America's most successful fintech firms have made investments in counterparts in India and Indonesia. Figure 11.6 shows the diffusion of business models, funds, and human capital in different parts of the world.

Since the early days of Naspers, SoftBank, and Rocket Internet, we have come a long way to this multidimensional, multifaceted global development of tech.

Global emerging markets are talking to each other much more these days

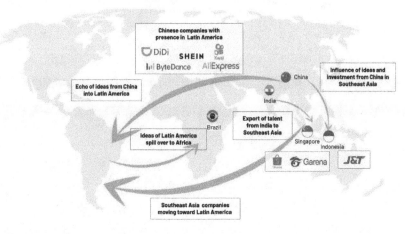

FIGURE 11.6 Global diffusion of business models from emerging markets
Source: © Momentum Works.

Or, as our good friend Dmitry Levit at Cento Ventures, a Singapore-based Venture Capital firm, has coined it: "the cross-pollination of global emerging markets."

We are still in the early stages of this pollination, and we do expect a few full blossoms down the road.

Epilogue

W E STARTED THIS BOOK with the view that there is a lot more going on with the waves of Chinese tech and internet companies and their expansion overseas than what is reported in the media or by analysts.

As our research and writing progressed, we realized that the experiences, lessons and reflections are very relevant to not only Chinese tech executives and those who have to work with, work for, compete against, invest in, or even regulate these expanding Chinese companies in their own countries, but also business leaders, entrepreneurs, investors, and executives in their own business and professional journey.

Leadership, people, organization, and product—in that order—are the key themes which define the success of any company.

Chinese tech businesses, being both Chinese and tech, offer a unique combination of experiences and practices that we can learn from, benchmark on, or try to avoid.

As we conclude this book, the US–China relationship has descended in a Cold War in all but name, and the whole world is entering uncharted waters: never has there been such deep integration of the economy and supply chain between the superpowers that are jostling for control and influence over so many fronts.

The development paths and strategies of West and East, instead of converging as hoped for in the 1990s and 2000s, have started to drift further and further apart. This brings a very challenging yet interesting phase for Chinese as well as global investors, entrepreneurs, and tech executives.

We take the optimistic view that some level of divergence and competition of development paths will eventually benefit humanity. Without a healthy level of competition, humanity as a whole will become less resilient in facing external challenges.

China was a powerful state throughout much of history—but in the eighteenth and nineteenth centuries it very quickly lost out to Europe, which was dynamic and also competitive. Why? Entire eras of prosperity generated rigidity, complacency, and unwillingness to change in the leadership, organization, and people. As a result, the product, which was the Chinese Empire, declined and eventually crumbled.

The lessons here are no different from how some large and successful corporations fail to adapt to changing environments, and no different from how our bodies build immune systems.

There will be setbacks, moments of frustration and even anger. In *Journey to the West*, if Tang Sanzang and his team had enjoyed a smooth journey for most of the parts, instead of going through 81 hardships one by one, they would have become so vulnerable that a single unexpected event might just wipe them out.

It's the same for Mao Zedong, the communist revolution leader. Although he was an avid student of history, he would not have formulated his strategic and military thoughts, and applied them to maximum effect if he had not suffered multiple setbacks in his early days. The mistakes he made in his old age, including the Cultural Revolution, precisely happened when seemingly nothing could challenge him anymore.

While complete Sino–US economic decoupling is unlikely to happen, we do not expect the world to return to the merry years of globalization, and the Sino–US jostle will be here for at least the next one to two decades. We need to be prepared for that, mitigating the risks but also seizing the opportunities that will arise.

Leaders, executives, and the rest of us should keep all this in mind, learn from history, and keep ourselves relevant and ahead of the game in this ever-changing world.

As Jack Ma's Alibaba always told its people: "Dream big, start small, learn fast."

Index